TURBULENCE AND RELATED PHENOMENA

Edited by **Régis Barillé**

Turbulence and Related Phenomena
http://dx.doi.org/10.5772/intechopen.74099
Edited by Régis Barillé

Contributors

Gilberto Javier Fochesatto, John Mayfield, Zambri Harun, Eslam Reda, Prabu K, Lu Lu, Zhiqiang Wang, Pengfei Zhang, Chunhong Qiao, Jinghui Zhang, Xiaoling Ji, Chengyu Fan, Regis Barille

Notice

Statements and opinions expressed in the chapters are these of the individual contributors and not necessarily those of the editors or publisher. No responsibility is accepted for the accuracy of information contained in the published chapters. The publisher assumes no responsibility for any damage or injury to persons or property arising out of the use of any materials, instructions, methods or ideas contained in the book.

First published in London, United Kingdom, 2019 by IntechOpen
IntechOpen is the global imprint of INTECHOPEN LIMITED, registered in England and Wales, registration number: 11086078, The Shard, 25th floor, 32 London Bridge Street
London, SE19SG – United Kingdom
Printed in Croatia

British Library Cataloguing-in-Publication Data
A catalogue record for this book is available from the British Library

Additional hard copies can be obtained from orders@intechopen.com

Turbulence and Related Phenomena, Edited by Régis Barillé
p. cm.
Print ISBN 978-1-83880-017-8
Online ISBN 978-1-83880-018-5

We are IntechOpen,
the world's leading publisher of
Open Access books
Built by scientists, for scientists

4,100+
Open access books available

116,000+
International authors and editors

120M+
Downloads

Our authors are among the

151
Countries delivered to

Top 1%
most cited scientists

12.2%
Contributors from top 500 universities

CLARIVATE ANALYTICS
BOOK
CITATION
INDEX
INDEXED

WEB OF SCIENCE™

Selection of our books indexed in the Book Citation Index
in Web of Science™ Core Collection (BKCI)

Interested in publishing with us?
Contact book.department@intechopen.com

Numbers displayed above are based on latest data collected.
For more information visit www.intechopen.com

Meet the editor

Régis Barillé received his PhD in electronics and Optronics in 1993 from the University of Montpellier. In 1994 he joined the University of Angers where he developed research in stimulated scatterings in nonlinear liquids and soliton propagation. In 1999 he was a visiting scientist at the University of Bordeaux. In 2002 he joined the VIRGO detector of gravitational waves project where he managed the injection bench. In 2007 he was visiting scientist at the Laser Physics Centre of the Australian National University. He is author or co-author of more than 120 publications in international periodicals and more than 50 conferences, mostly invited. He has a great deal of experience in interdisciplinary work ranging from experimental physics and environmental metrology to the characterization of surfaces.

Contents

Preface

The importance of atmospheric turbulence has increased in recent years in relation to different topics, such as understanding the turbulence layer for communications. Laser beam propagation can create a large number of optical link applications which can lead to the creation of such free-space optical (FSO) communications.

The fluctuations in the index of refraction of the atmosphere result in many optical effects long known to astronomers. Some of these multiple effects—such as twinkling, quivering, tremor disc, dancing, wandering, breathing, image distortion, and boiling—still require complete description and modelization. While previous applications have focused on the characterization of atmospheric turbulence limiting the laser propagation, the advent of free-space-enabled communications requires novel methods for secure and robust links in response to a growing need for high-speed and "free from electromagnetic interference" communication systems. Links between satellites, ground stations, unmanned aerial vehicles, high buildings in the city for local and metropolitan area networks, and other nomadic communication partners are of practical interest.

In many cases, this has also led to the development of other related areas of research, such as underwater optical propagation. An important characteristic of this area of atmospheric turbulence is that it has been explored by multiple communities such as astronomy, optical metrology, and environmental analysis. In many cases, these communities tend to have some overlap, but are largely disjointed and carry on their research independently.

One of the goals of this book is to bring together researchers of different communities in order to maximize the cross-disciplinary understanding of this area. Another aspect of the subject of atmospheric turbulence is that there seems to be a distinct set of researchers working on newer aspects of atmospheric turbulence in the context of emerging research into, for example, urban atmospheric layers, urban metrology, probing the atmosphere, and measuring fluid flow.

This book is also an attempt to discuss both the current and modern aspects of FSO communication in a unified way. Chapters are devoted to various techniques to enhance FSO communication system performance. Other chapters are concerned with atmospheric turbulence measurements.

In addition, the book also demonstrates different aspects of FSO communications in the context of modern applications in information networks. Many new results, such as turbulent flows in industrial applications and atmospheric phenomena, have also been explored for the first time in this book.

Each chapter in the book is structured as a comprehensive survey which discusses the key models and results for the particular area. In addition the future trends and research directions are presented in each chapter. It is hoped that this book will provide a comprehensive understanding of the area for students, professors, and researchers.

I would like to acknowledge all the colleagues who assisted me in the making of this book with their fruitful discussions.

Régis Barillé
Laboratory MOLTECH-Anjou
University of Angers, France

Chapter 1

Introductory Chapter: Turbulence and Related Phenomena

Régis Barillé

Additional information is available at the end of the chapter

http://dx.doi.org/10.5772/intechopen.83378

1. Introduction

This book presents some of the last results concerning the atmospheric turbulence and all its effect on the propagation of light beam. This domain of research covers some related applications to optical communications and the understanding of new optical effects due to atmospheric conditions.

The process of optical beam propagation through random media has been studied for many years and has followed the development of the laser technology. The term random medium or turbulent medium means that the index of refraction of the medium exhibits random spatial variations with dimensions larger than the optical wavelength of the propagating optical beam. Fluctuations in the refractive index in air, i.e., air density, are caused mainly by temperature variations affecting the wave front of a light beam. Atmospheric refractions can cause spatial and temporal (intensity) variations in propagating beams. The fluctuations in the index of refraction of the earth's atmosphere result in many optical effects well-known to astronomers. Some of these effects are twinkling (variation of image brightness), quivering (displacement of image from normal position), smearing of the diffraction image, wandering (continuous movement of a star image about a mean point), wandering (slow oscillatory motions of the image for a period of approximately 1 minute and angular excursions of a few seconds of arc), pulsation (fairly rapid change of size of the image), image distortion, and boiling (time-varying nonuniform illumination in a larger spot image) [1, 2].

The first studies concerning the propagation of unlimited plane waves and spherical waves through random media led to the classical books published in the early 1960s by Tatarskii, *Wave Propagation in a Turbulent Medium*, and LA Chernov, *Wave Propagation in a Random Medium*. VI Tatarskii predicted based on his theory considering weak fluctuations that the correlation width of the irradiance fluctuations is on the order of the first Fresnel zone L/k where L is the distance to the source and k is the wave number. Their results concerning optical scintillation were only limited to weak fluctuations. However, in the case of high turbulence, saturation effects of optical waves occur and were first experimentally demonstrated later by ME Gracheva and AS Gurvich in 1965. The publication of this work stimulated a lot of theoretical and experimental studies related to irradiance fluctuations under conditions of strong turbulence. So, the base for the development of studies concerning turbulence effects on laser beam in atmospheric turbulence was set down. Later in the aim to better improvements of the theoretical bases of the saturation phenomenon, several qualitative models describing the underlying physics associated with amplitude or irradiance fluctuations were developed in the mid-1970s. A generalization of the Tatarskii's physical optics model was published where the loss of spatial coherence of the wave as it propagates into the strong fluctuation regime was included.

Until now, there are no easy solutions to deal the problem of irradiance fluctuations that applies to all conditions of optical turbulence for the propagation of electromagnetic waves. This problem recently gains importance since free space optical communications (FSO) is now common for point-to-point communications not only between fixed locations on land but also for communication between moving platforms like vehicles on land, on the surface of the sea, in air, and in space [3]. Free space communications imply that it is not practical or impossible to use optical fiber technology to connect the points that need to communicate or exchange data. Since the beginning of the twenty-first century, there is a growing interest in increasing the capacity of free space telecommunication systems to eventually the pending bandwidth limitation. The current development of new protocols in free-space optical communication requires the knowledge of light beam propagation in a turbulent medium and the possibility to change the laser beam parameters (beam shape, coherence, etc.). FSO is a technology that uses the visible and infrared light propagating through the atmospheric medium to transmit information. However, the constitution of the atmosphere involves turbulence, particularly aerosols (fog, smoke, and dust) have similar particle size distributions compared with optical wavelengths in FSO. In another domain of activities atmospheric turbulence induces spatial and temporal changes during the light propagation and in transmitted signals acquired by the receptors. In turn, these effects can significantly degrade (blur, shimmer, and distort) optical data. This can also potentially result in scattering and absorption of visible and IR optical beams. Even in clear outdoor conditions, wireless optical links experience fluctuations in both the intensity and the phase of an optical wave propagation.

These conditions lead to degradation of the FSO link performances and its availability for a large development in particular in an urban environment where the conditions of propagation are not clearly understood. Moreover, in some cases, the atmospheric parameters can impact the beam propagation and give insights of the turbulence phenomena involved. For all these reasons, it is important continuing to develop studies on light beam in atmospheric turbulence.

Author details

Régis Barillé

Address all correspondence to: regis.barille@univ-angers.fr

MOLTECH-Anjou, University of Angers/UMR CNRS 6200, Angers, France

References

[1] Andrews LC, Phillips RL, Hopen CY. Laser Beam Scintillation with Applications. Bellingham: SPIE Optical Engineering Press; 1998

[2] Murty SSR. Laser beam propagation in atmospheric turbulence. Proceedings of the Indian Academy of Sciences Section C: Engineering Sciences. 1979;**2**(Part. 2):179-195

[3] Malik A, Singh P. Free space optics: Current applications and future challenges. International Journal of Optics. 2015;**2015**:945483

Performance Analysis of FSO Systems over Atmospheric Turbulence Channel for Indian Weather Conditions

Prabu Krishnan

Additional information is available at the end of the chapter

http://dx.doi.org/10.5772/intechopen.80275

Abstract

Free-space optical (FSO) communication is a line of sight (LOS) technology and has significant advantages and attractive applications. Recently, spectrum slicing wavelength division multiplexing (SS-WDM)-based FSO systems provide improved link range, high capacity, and efficiency. In this chapter, the SS-WDM-based FSO system is proposed with four channels to increase the performance of communication under various wind speed and heights of the buildings. But, atmospheric turbulence fading, scintillation, and pointing errors (PE) are the main impairments affecting the performance of FSO communication systems. Predominantly, the turbulence variation due to wind velocity, refractive index, and height of buildings has been majorly focused and analyzed for Vellore weather conditions. A case study has been experimented on how the height of buildings and the atmosphere around VIT, Vellore campus, affect the transmission of light in free space. The bit error rate of the proposed system is analyzed with distance, received power for various wind speed and different heights of the buildings.

Keywords: free space optics, spectrum slicing wavelength division multiplexing (SS-WDM), bit error rate, atmospheric turbulence

1. Introduction

Free space optics (FSO) is an emerging and promising technology for next-generation wireless communication applications like short-range indoor wireless communication, back-haul for wireless cellular networks, last mile access, high-definition television (HDTV) transmission, and laser communications in space. In comparison with traditional radio frequency communications, the attractive features in the FSO communications include license-free operation, simple deployment, high data rate, and high transmission security [1].

However, the performance of FSO communication systems is extremely dependent of the atmospheric weather conditions. When the atmospheric channel conditions are poor, then a transmitted light signal is affected by scattering, absorption, and turbulence. The inhomogeneity in the temperature, pressure, and wind speed over the channel varies the refractive index of the atmosphere and it creates the optical signal intensity fluctuation. The negative impacts of turbulence include scintillations, phase-front distortions, beam spreading, and beam wander [2, 3]. Another significant problem with FSO links is that they are relying on the pointing performance. Errors in tracking systems, mechanical misalignment, and vibrations of the transmitter beam due to building sway phenomena lead to further performance degradation as a result of pointing errors (PE) [4].

The effects of atmospheric turbulence can be mitigated by performing aperture averaging or employing diversity at the receiver. Introducing multiple apertures at the transmitter/or the receiver provides the multiple-input multiple-output (MIMO) FSO systems potentially have to enhance performance of the system. The various forms of diversity schemes are temporal, spatial, and wavelength [5]. Relay-assisted communication is the alternate way to reduce the effects of turbulences [6]. In order to overcome this issue, certain techniques like SS-WDM were implemented. It offers high spectral efficiency and a wide coverage area which facilitates more number of users. It is a scalable network which marks its specialty in the optical networking communication area [7].

Spectrum slicing technique is used to modulate the optical signals. The desired spectrum is being sliced differently in order to modulate the optical signals accordingly and analyze the spectrum with the respective parameters [8]. Parallel transmission of slicing from a single broadband noise source has highest potential for creating a multichannel system. Spectrum slicing has a higher potential for future fiber to home access network which can further be incorporated in any optical system where low power consumption is preferred [9].

2. System model

The transmitter section comprises pseudo-random bit sequence (PRBS) generator, NRZ pulse generator, followed by Mach-Zehnder modulator (MZM) and a continuous wave laser, whereas the receiver consists an APD photodiode and a low-pass Bessel filter. Performance of the communication link is inspected using BER analyzer. PRBS generates information signal in the form of binary pulses, i.e., 1010101 and so on which are transformed to electrical signal thereby directing toward NRZ pulse generator. The conversion of binary pulses to electrical signals occurs (**Figure 1**). The WDM FSO link optimizing in this study was used and modified as details given in [10].

The output of NRZ pulse generator is given to MZM whose other input end receives input from a continuous wave laser. It converts the electrical signal to an optical signal. The signal is now boosted using an optical amplifier directly without converting to an electrical signal. The signal heads through the FSO channel in the form of narrow-beam electromagnetic wave which is received by a photodetector, converting optical to its corresponding electrical signal. Further,

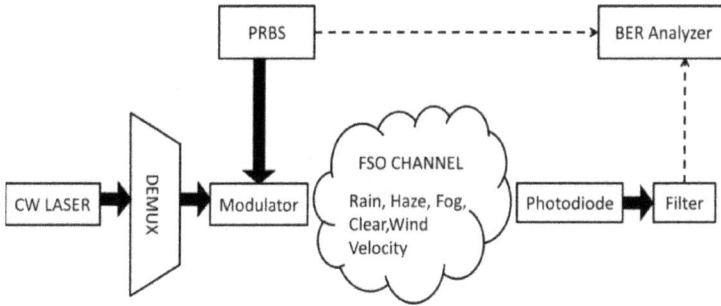

Figure 1. Architecture of SS-WDM-FSO communication system.

a low-pass Bessel filter suppresses the noise which is a dominant part of the signal. Desired output is achieved at the receiver which is visualized and inspected using BER analyzer [10].

3. Channel model

The atmospheric attenuation and turbulence are the major challenges in FSO communication system.

3.1. Atmospheric attenuation

The atmospheric attenuation loss modeled by Beers-Lambert law is given as [11]:

$$h_l = e^{-\sigma L} \tag{1}$$

where σ denotes a wavelength and weather-dependent attenuation coefficient and L is the propagation distance.

3.2. Atmospheric turbulence

The atmospheric turbulence is classified as weak, moderate, strong, and saturated regimes based on the variation of refractive index and inhomogeneity. The different mathematical models are developed to represent the turbulence regimes like log-normal, negative exponential, and gamma-gamma and M-distribution to represent weak, strong, weak-to-strong, and generalized turbulences, respectively [12–14]. The atmospheric turbulence models describe the probability density function statistics of the irradiance fluctuation.

3.2.1. Lognormal

In this model, the statistics of the irradiance fluctuations obeys the log-normal distribution. This model is characterized by a single scattering event and is best suited for weak turbulence regime. The PDF can be given as [15, 16]:

$$f(I) = \frac{1}{\sqrt{2\pi\sigma_I^2}} \frac{1}{I} \exp\left(-\frac{\left(\ln\left(\frac{I}{I_0}\right) - E[I]\right)^2}{2\sigma_I^2}\right), I \geq 0 \qquad (2)$$

where σ_I^2 is the Rytov varience, I be the field irradiance in turbulent medium while the intensity in free-space (no turbulence) is represented as I_0, and E[I] is the mean log intensity.

3.2.2. Negative exponential

In this model, the number of independent scattering is very high and it can support for saturation regime. Therefore, the irradiance fluctuation follows the Rayleigh distribution entailing negative exponential statistics for the irradiance. The negative exponential PDF can be given as [15, 16]:

$$f(I) = \frac{1}{I_0} \exp\left(-\frac{I}{I_0}\right), I_0 > 0 \qquad (3)$$

where $E[I] = I_0$ is the mean received irradiance.

3.2.3. Gamma-gamma

The atmospheric turbulence is modeled by gamma-gamma distribution with scintillation parameters α and β, which are indicated as functions of the Rytov variance and a geometry factor. The PDF of the gamma-gamma channel model is given by [11]:

$$f_{I_s}(I_s) = \frac{2(\alpha\beta)^{(\alpha+\beta)/2}}{\Gamma(\alpha)\Gamma(\beta)} I_s^{(\alpha+\beta)/2-1} K_{(\alpha-\beta)}\left(2\sqrt{\alpha\beta I_s}\right) \qquad (4)$$

where α and β are the effective number of large- and small-scale turbulent eddies, $\Gamma(\cdot)$ is the gamma function, and $K_{(\alpha-\beta)}$ is the modified Bessel function of the second kind of order $(\alpha - \beta)$. The effective number of large- and small-scale turbulent eddies α and β for a spherical wave is given by [11]:

$$\alpha = \left[\exp\left(\frac{0.49\delta_n^2}{\left(1 + 0.18d^2 + 0.56\delta_n^{12/15}\right)^{7/6}}\right) - 1\right]^{-1} \qquad (5)$$

$$\beta = \left[\exp\left(\frac{0.51\delta_n^2\left(1 + 0.69\delta_n^{12/15}\right)^{-5/6}}{\left(1 + 0.9d^2 + 0.62d^2\delta_n^{12/15}\right)^{5/6}}\right) - 1\right]^{-1} \qquad (6)$$

where $d = \sqrt{kD^2/4L}$, $k = 2\pi/\lambda$, is the optical wave number with λ being the operational wavelength, L is the length of the optical link, and D is the receiver's aperture diameter. The parameter

δ_n^2 is the Rytov variance and is given as: $\delta_n^2 = 0.5C_n^2 k^{7/6} L^{11/6}$ and the C_n^2 represents the refractive index structure parameter. This model is valid for all turbulence regimes from weak to strong and the gamma-gamma distribution approaches negative exponential distribution when it approaches saturation regime [17, 18].

3.2.4. M-distribution

The transmitted signal is scattered due to natural turbulences such as rain, fog, smoke, smog, and heavy dust particles in the atmospheric channel (AC). The combined effects of fading due to atmospheric turbulence and misalignment are considered and the combined unconditional probability density function (PDF) for M-distribution has been derived in [19]. As derived in [19], the PDF of the irradiance h is given by:

$$f_I(I) = \frac{g^2 A}{2I} \sum_{k=1}^{\beta} a_k \left(\frac{\alpha\beta}{\mu\beta + \Omega'} \right)^{-\frac{\alpha+k}{2}} G_{1,3}^{3,0} \left[\frac{\alpha\beta}{\mu\beta + \Omega'} \frac{I}{A_0} \Big|_{g^2, \alpha, k}^{g^2+1} \right] \tag{7}$$

where $\mu = 2b_0(1 - \rho)$ is a large-scale scattering parameter, β is the quantity of fading parameter, A_0 is the fraction of the collected power at r = 0 (radial distance), $\Omega' = \Omega + 2\rho b_0 + 2\sqrt{2\rho b_0 \Omega} \cos(\phi_A - \phi_B)$ be the average power. The amount of scattering power coupled to the LOS component is denoted by the parameter ρ and its range from 0 to 1. The parameters ϕ_A and ϕ_B are the deterministic phases of the LOS and the coupled-to-LOS component. The parameters g, A and a_k are defined as [19]

$$g = \frac{w_{zeq}}{2\sigma_s} \tag{8}$$

$$A = \frac{2\alpha^{\alpha/2}}{\mu^{1+\alpha/2}\Gamma(\alpha)} \left(\frac{\mu\beta}{\mu\beta + \Omega'} \right)^{\beta+\alpha/2} \tag{9}$$

$$a_k = \binom{\beta-1}{k-1} \frac{(\mu\beta+\Omega')^{1-0.5k}}{(k-1)!} \left(\frac{\Omega'}{\mu} \right)^{k-1} \left(\frac{\alpha}{\beta} \right)^{0.5k} \tag{10}$$

where σ_s is the pointing error displacement standard deviation at the receiver, w_{zeq} be the equivalent beam radius and can be calculated by using the relations $v = \sqrt{\pi} a/\sqrt{2} w_z$, $w_{zeq} = w_z^2 \sqrt{\pi} \text{erf}(v)/2ve^{-v^2}$, where w_z is the beam waist at distance z, a is the radius of a circular detection aperturture. The gamma-gamma, K-distribution, and negative exponential model are obtained by the values of the parameter$(\rho = 1, \Omega' = 1)$, $(\rho = 0, \Omega = 0 \text{ or } \beta = 1)$, and $(\rho = 0, \Omega = 0 \text{ or } \alpha \rightarrow \infty)$, respectively.

3.3. Wind speed-induced turbulence

Wind control is a standout among the most essential wellsprings of feasible vitality which is sustained on a little and additionally extensive scale. It is the most noticeable factor which weakens the optical signal which propagates over free space [20]. The optical signal blurs as the climatic visibility diminishes because of the turbulence which relates the refractive record nonhomogeneity parameter of the particles introduced in the air. In Ref. [21], the authors experimented the FSO link which is located at Milesovka hill situated 70 km north to Prague. It is been seen that flag lessening is conversely relative to meteorological visibility. The visibility at the transmitter and receiver sites is interfered by the wind speed which can be mathematically modeled using 3D vector where refractive index and rain rate are continuously recorded which estimate quantitative impact on attenuation.

Atmospheric turbulence brings about irregular change of the refractive index of the light propagation path. This refractive index change is the immediate final result of arbitrary discrepancies in atmospheric temperature from point to point [22]. These random temperature fluctuations are a function of the atmospheric pressure, altitude, and wind speed.

Wind is also persuading the received power of the FSO link signal. Wind, especially wind turbulences, causes hurdle changes of the atmosphere refractive index which is redistributing the optical beam of the FSO link. In order to investigate the wind influence on the FSO link attenuation, we selected important wind parameters influencing the FSO link attenuation due to this optical energy redistribution.

Therefore, the temperature and wind speeds are measured and the attenuation due to wind turbulence is calculated. Based on the calculated attenuation, the BER versus received power and distance is analyzed for various heights of buildings and minimum, maximum wind speeds.

The relative humidity, minimum, and maximum temperature variation during April 2017 at Vellore 12.92°N/79.13°E, 267 m is shown in **Figure 2**. The minimum and maximum temperature obtained in the month of April 2017 is 21 and 43°C. **Figure 3** illustrates the speed and direction of wind during April 2017 at Vellore, India. The direction of wind is represented using angles $0, 90, 270$, and $360°$ for South, West, North, and East, respectively. The minimum and maximum speed of wind recorded in the month of April 2017 is 2.5 and 26 km/h.

The atmospheric turbulence because of wind speed introduces radical fluctuations in refractive index that affects the propagation FSO signal [23]. In order to analyze the influence of wind on

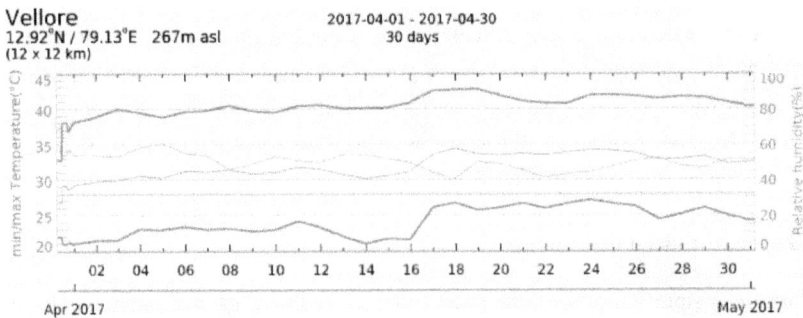

Figure 2. Temperature and relative humidity during April 2017 at Vellore, India [23].

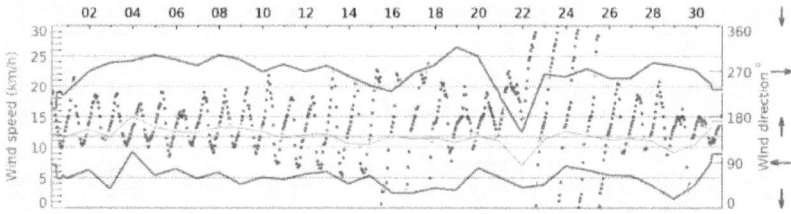

Figure 3. Wind speed and direction during April 2017 at Vellore, India [23].

FSO system, the turbulent energy, direction, and velocity of the wind have to be considered. The turbulent energy of the wind at every direction can be calculated as:

$$E_t = \frac{1}{2N} \sum \left[(x - \bar{x})^2 + (y - \bar{y})^2 + (z - \bar{z})^2 \right]$$ (11)

where $\bar{x}, \bar{y}, \bar{z}$ are the given wind speeds in a particular direction, x, y, z are the cumulative wind speed in any one direction, N be the number of samples, and E_t is the turbulent energy. The turbulent energy represents wind velocity standard deviation. The attenuation caused by the turbulence can be calculated by a regressive formula which is as follows:

$$A = 70 - 73e^{-0.2867Et}$$ (12)

4. Results and discussion

The BER of the proposed system is analyzed with respect to transmission distance and received power over various wind speeds and different heights of buildings. The results are compared with and without spectrum slicing-based WDM FSO system. It shows the importance of spectrum slicing in WDM-based FSO systems.

This imparts that wind speed in any direction can be correlated to FSO link attenuation. Since the wind speed affects the propagation of the signal, the height of the buildings also obstructs the signal transmission. The analysis on how the heights of buildings play a significant role is deeply inspected in VIT University, Vellore, Tamil Nadu, India. The wind speed velocities have been captured during the month of April 2017. The current status of the wind has viewed from meteorological media which gives the flow of wind in every direction. This assists in calculating the exact direction of wind and has been analyzed using Eqs. (11) and (12) respectively. The attenuation values have been calculated for maximum and minimum values of the wind speed. The buildings in VIT for which it has been experimented are Technology Tower (TT), Silver Jubilee Tower (SJT), GDN, and Rajeswari Tower, which have the range of heights from 10 to 50 m. The graphs have been plotted in order to analyze the effect of turbulence on the FSO link with building heights of 10 and 80 m. In Vellore region, July to September is the monsoon season. So, we have considered the rain data from July to September.

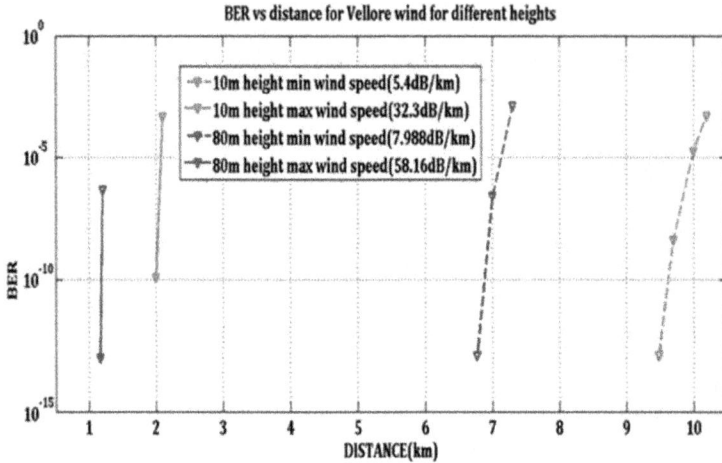

Figure 4. BER in terms of distance for Vellore weather conditions affected by wind velocity and building heights.

The BER in terms of transmission distance for four-channel SS-WDM FSO system is illustrated in **Figure 4**. The error rate performance is analyzed with respect to various building heights, minimum and maximum wind speed. From the figure, it is observed that at a height of 10 m, the BER of 10^{-10} is achieved at 9 km distance under minimum wind speed. But, for the same height the same BER can be achieved only at the link range of 2 km under maximum wind speed. That is, the wind speed decreases the link range about 7 km from the analysis. At 80 m height, the minimum BER of 10^{-13} is achieved at 7 and 1 km over minimum and maximum wind speed, respectively. It is inferred that at both heights, the speed of wind decreases the link range of an FSO system.

The BER versus received power shown in **Figure 5** is the received power analysis done in Vellore, Tamil Nadu, India. It shows how the received power of an FSO link is affected by building heights at various wind velocities. As we expected that the received power increases, the BER decreases. From the figure, it is observed that at a height of 10 m, near the received power of −70 dBm under influence of minimum wind speed, the BER recorded is 10^{-9}. Near the received power of −72 dBm, spectrum slicing gives a qualitatively better BER value of 10^{-6} when the building height is 80 m for maximum wind speed.

In forest area due to high evapotranspiration, that is, the water from the plants, trees, and land surface are transferred to the atmosphere in the form of gas, optical signal fading will be high. Because, the phase changed water contents (gas) use to absorb the energy and cool the land surface and the end result is a high level of turbulence compared with urban areas. Wind direction can be imagined as a horizontal flow of numerous rotating eddies, that is, turbulent vortices of various sizes, with each eddy having horizontal and vertical components [24] is shown in **Figure 6**. Since the turbulence level is high, the transmitted optical signal scintillation will be more with the effects of high BER.

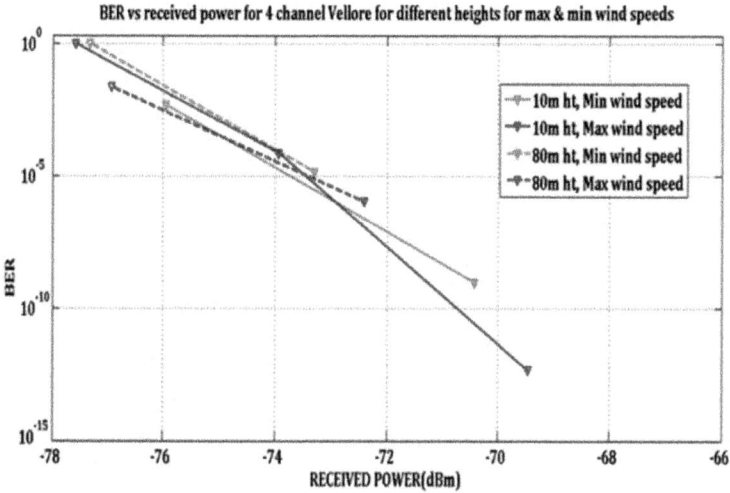

Figure 5. BER versus received power for Vellore weather conditions affected by wind velocity and building heights.

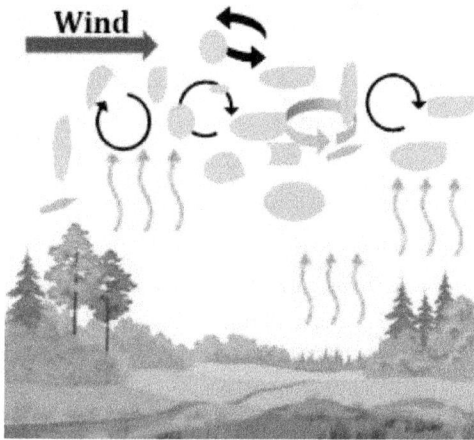

Figure 6. Rotation of turbulent eddies with respect to horizontal wind flow.

Figure 7 demonstrates the BER versus distance plot for forests under the effect of wind speed at various heights of 15, 33, and 97 m for four-channel system model. At 15 m height, it is noticed that in case of SS-WDM-FSO, BER value is 10^{-10} at a distance of 210 km. In case of WDM-FSO, the BER is 10^{-8} at the same distance. For a forest height of 33 m, it is seen that a BER value of 10^{-6} is achieved at a distance of 10 km, when no spectrum slicing is done on the system, whereas for the same distance, BER value of nearly 10^{-7} is obtained when the FSO channel is subjected to slicing. So, it can be said that slicing the channel does give a considerable amount of change in BER values. For forest with tree heights around 97 m under influence

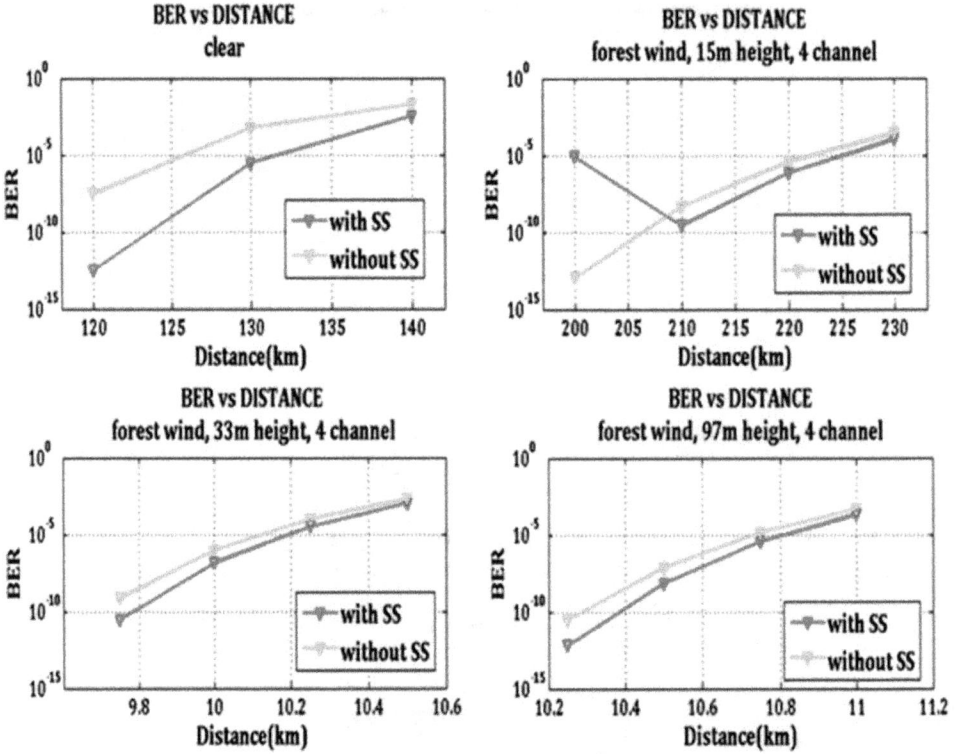

Figure 7. BER in terms of distance for forests under influence of wind speeds at various heights of 15, 33, and 97 m for four-channel model.

of wind, it is observed that the BER value of 10^{-10} is obtained at an FSO link length of 10.4 km and 10^{-8} at the same distance with and without slicing, respectively. This shows that a low BER value is obtained in case of spectrum slicing.

5. Conclusion

In this chapter, the SS-WDM FSO system is proposed and the importance of spectrum slicing on WDM FSO systems is analyzed in terms of average BER. Effect of wind velocity as well as turbulent energy on various building in VIT University has been deeply studied and results have been plotted. The performance of the SS-WDM-FSO system is analyzed for various building heights, wind speeds in terms of distance and received power. It has been observed that the wind speed and height of buildings decreases the link range of FSO system. It also affects the BER performance of the system. Also observed is that the spectrum slicing reduced the number of components and losses. It improves the spectrum efficiency of the system, and it is compact and cost effective.

Author details

Prabu Krishnan

Address all correspondence to: nitprabu@gmail.com

Department of Electronics and Communication Engineering, National Institute of Technology Karnataka, Surathkal, India

References

[1] Peppas KP, Stassinakis AN, Topalis GK, Nistazakis HE, Tombras GS. Average capacity of optical wireless communication systems over IK atmospheric turbulence channels. 2012; 4(12):1026-1032

[2] Krishnan P, Kumar DS. Performance analysis of free-space optical systems employing binary polarization shift keying signaling over gamma-gamma channel with pointing errors. Optical Engineering. 2014;53(7):076105-076105

[3] Prabu K, Sriram Kumar D. MIMO free-space optical communication employing coherent BPOLSK modulation in atmospheric optical turbulence channel with pointing errors. 2015;343:188

[4] Sandalidis HG. Coded free-space optical links over strong turbulence and misalignment fading channels. IEEE Transactions on Communications. 2011;59(3):669-674

[5] Bayaki E, Schober R, Mallik R. Performance analysis of MIMO free-space optical systems in gamma–gamma fading. IEEE Transactions on Communications. 2009;57(11):3415-3424

[6] Laneman JN, Tse DNC, Wornell GW. Cooperative diversity in wireless networks: Efficient protocols and outage behavior. 2004;50(12):3062-3080

[7] Prabu K, Sriram Kumar D. BER analysis of DPSK–SIM over MIMO free space optical links with misalignment. Optik. 2014;125(18):5176-5180

[8] Aladeloba AO, Woolfson MS, Phillips AJ. WDM FSO network with turbulence-accentuated interchannel crosstalk. 2013;5(6):641-651

[9] Ciaramella E, Arimoto Y, Contestabile G, Presi M, D'Errico A, Guarino V, Matsumoto M. 1.28 Terabit/s (32 × 40 Gbit/s) WDM transmission system for free space optical communi-cations. IEEE Journal on Selected Areas in Communications. 2009;27(9):1639-1645

[10] Prabu K, Charanya S, Jain M, Guha D. BER analysis of SS-WDM based FSO system for Vellore weather conditions. Optics Communications. 2017;403:73-80

[11] Andrews LC, Phillips RL. Laser Beam Propagation through Random Media. Bellingham: SPIE Press; 2005

[12] Killinger D. Free space optics for laser communication through the air. 2002;13:36-42

[13] Nistazakis HE, Karagianni EA, Tsigopoulos AD, Fafalios ME, Tombras GS. Average capacity of optical wireless communication systems over atmospheric turbulence channels. 2009; **27**:974-979

[14] Nistazakis HE, Tsiftsis TA, Tombras GS. Performance analysis of free-space optical communication systems over atmospheric turbulence channels. IET Communications. 2009;**3**: 1402-1409

[15] Ghassemlooy Z, Popoola W, Rajbhandari S. Optical Wireless Communications System and Channel Modelling with MATLAB. Boca Raton, Florida's: CRC Press; 2012

[16] Ghassemlooy Z, Tang X, Rajbhandari S. Experimental investigation of polarisation modulated free space optical communication with direct detection in a turbulence channel. IET Communications. 2012;**6**:1489-1494

[17] Garrido-Balsells JM, Jurado-Navas A, Paris JF, Castillo-Vázquez M, Puerta-Notario A. On the capacity of $\mathcal{M}$-distributed atmospheric optical channels. Optics Letters. 2013;**38**(20): 3984-3987

[18] Tang X, Xu Z, Ghassemlooy Z. Coherent polarization modulated transmission through MIMO atmospheric optical turbulence channel. 2013;**31**(20):3221-3228

[19] Yang L, Hasna MO, Gao X. Asymptotic BER analysis of FSO with multiple receive apertures over $\mathcal{M}$-distributed turbulence channels with pointing errors. Optics Express. 2014;**22**(15):18238-18245

[20] Turbulence Intensity in Complex Environments and its Influence on Small Wind Turbines. Uppsala, Sweden: Nicole Carpman; 2011

[21] Fiser O et al. FSO link attenuation measurement and modelling on Milesovka hill. In: Proceedings of the 5th European Conference on Antennas and Propagation (EUCAP). Rome, Italy: IEEE; 2011

[22] Pratt WK. Laser Communication Systems. 1st ed. New York: John Wiley & Sons, Inc.; 1969

[23] https://www.meteoblue.com/en/weather/forecast/week/vellore_india_1253286 (Data for wind speed)

[24] Wang K, Dickinson RE. A review of global terrestrial evapotranspiration: Observation, modeling, climatology, and climatic variability. Reviews of Geophysics. 2012;**50**(2):1-54

Chapter 3

Laser Beam Propagation through Oceanic Turbulence

Zhiqiang Wang, Lu Lu, Pengfei Zhang,
Chunhong Qiao, Jinghui Zhang, Chengyu Fan and
Xiaoling Ji

Additional information is available at the end of the chapter

http://dx.doi.org/10.5772/intechopen.76894

Abstract

Using a recently proposed model for the refractive index fluctuations in oceanic turbulence, optical beam propagation through seawater is explored. The model provides an accurate depiction of the ocean through the inclusion of both temperature and salinity fluctuations to the refractive index. Several important statistical characteristics are explored including spatial coherence radius, angle-of-arrival fluctuations, and beam wander. Theoretical values of these parameters are found based on weak fluctuation theory using the Rytov method. The results presented serve as a foundation for the study of optical beam propagation in oceanic turbulence, which may provide an important support for further researches in applications for underwater communicating, imaging, and sensing systems.

Keywords: oceanic turbulence, laser beam, spatial coherence radius, angle-of-arrival fluctuations, beam wander

1. Introduction

The study of optical wave propagation through random media is a perpetually important topic for its many applications in the atmosphere and the ocean. Random fluctuations in the index of refraction cause beam spreading (beyond that due to pure diffraction), loss of spatial coherence, random wandering of the instantaneous beam center, and random fluctuations in the irradiance and phase [1]. The index of refraction fluctuations, generally referred to as optical turbulence, is one of the most significant quantities in optical wave propagation. For different random media, there are some differentiations among the index of refraction fluctuations. The index of refraction of atmosphere is primarily caused by fluctuating temperature. The refraction index in seawater is induced not only by temperature fluctuations but also by fluctuations of salinity. Changes in the optical signal due to absorption or scattering by molecules or particles are not considered here. Under the assumption of a statistically homogeneous and isotropic ocean, the power spectrum of oceanic turbulence is determined by fluctuations of refraction index.

With the development of underwater optical communications, imaging, sensor, and laser radar, it is indispensable to investigate the propagation behavior of laser beams through water medium. Knowledge of beam spreading is extremely important in a free space optics (FSO) communications link because it determines the loss of power at the receiver. The spatial coherence radius defines the effective receiver aperture size in a heterodyne detection system [1]. To the coherence degradation of laser beams, the spatial coherence radius can also be described as the strength of oceanic turbulence. Angle-of-arrival (AOA) fluctuations of an optical wave in the plane of the receiver aperture are associated with image jitter (dancing) in the focal plane of an imaging system [1] so that it plays a critical role in beam wave propagation applications such as imaging, lasercom, and other related areas. Movement of the short-term beam instantaneous center (or "hot spot") is commonly called beam wander [1]. Beam wander is an important propagation characteristic of laser beams, which determines their utility for practical applications, such as laser communication [2, 3] and global quantum communication [4].

In this chapter, Section 2 describes a brief introduction of oceanic turbulence including the power spectrum and several significant oceanic parameters. The spatial coherence radius of a plane wave and a spherical wave propagating through oceanic turbulence has been investigated in Section 3, which are valid in both weak and strong fluctuations. Section 4 describes the angle-of-arrival fluctuations for plane- and spherical-wave models of oceanic turbulence. Based on the oceanic power spectrum, the beam wander effect with analytical and numerical methods in weak fluctuation theory is shown in Section 5. These results may provide an inroad for understanding laser beam propagation through oceanic turbulence, and the theoretical findings may provide an important support for further researches in applications for underwater communicating, imaging, and sensing systems.

2. Nature of oceanic turbulence

Turbulence is a random, three-dimensional motion with the velocity and vorticity irregularly distributed in time and space [5]. In general, turbulence is accepted to be an energetic, rotational, and eddying state of motion that results in the dispersion of material and the transfer of momentum, heat, and solutes at rates far higher than those of molecular processes alone [6]. It is characterized by an energy transfer from large to small scales where the dissipation of kinetic energy is taking place [5]. Oceanic motions are constrained to flow along density surfaces by the Earth's rotation and the density stratification. In the upper ocean, microscale turbulence is generated by surface winds, air-sea cooling, or evaporation. In the ocean interior, microscale turbulence develops when internal waves develop strong shears and overturn and break, much like surface gravity waves [7]. These breaking events play a fundamental role in the ocean circulation, because they mix the densest waters at the ocean bottom with the lighter waters above, thereby allowing the densest waters to come back to the surface [7]. Much of the turbulence induced in benthic boundary layers is driven by external processes resulting from the fluxes of buoyancy and momentum through the nearby boundary, such as a tidally driven current, geothermal heat flux, and so on [6]. They are driven by sources of energy outside the benthic boundary layer itself [6].

2.1. Power spectrum of oceanic turbulence

Since the power spectrum of oceanic turbulence proposed in 2000 [8], there has been remarkable interest in the study of propagation characteristics using laser beams in seawater. The power spectrum of oceanic turbulence has been simplified for homogeneous and isotropic water media [9], which is applicable for isothermal water [10]. When the eddy thermal diffusivity and the diffusion of salt are assumed to be equal, the power spectrum for homogeneous and isotropic oceanic water is given by the expression [11].

$$\Phi_n(\kappa) = 0.388 \times 10^{-8} \varepsilon^{-1/3} \kappa^{-11/3} \left[1 + 2.35(\kappa\eta)^{2/3}\right] \frac{\chi_T}{w^2} \left(w^2 e^{-A_T\delta} + e^{-A_S\delta} - 2we^{-A_{TS}\delta}\right), \quad (1)$$

where ε is the rate of dissipation of kinetic energy per unit mass of fluid ranging from about $10^{-10} \mathrm{m}^2/\mathrm{s}^3$ in the abyssal ocean to $10^{-1} \mathrm{m}^2/\mathrm{s}^3$ in the most actively turbulent regions. χ_T is the rate of dissipation of mean-squared temperature and has the range $10^{-4} \mathrm{K}^2/\mathrm{s} - 10^{-10} \mathrm{K}^2/\mathrm{s}$, w defines the ratio of temperature and salinity contributions to the refractive index spectrum, which varies in the interval $[-5, 0]$, with -5 and 0 corresponding to dominating temperature-induced and salinity-induced optical turbulences, respectively [11]. In addition, η is the Kolmogorov microscale (inner scale), and $A_T = 1.863 \times 10^{-2}$, $A_S = 1.9 \times 10^{-4}$, $A_{TS} = 9.41 \times 10^{-3}$, $\delta = 8.284(\kappa\eta)^{4/3} + 12.978(\kappa\eta)^2$ [11].

2.2. Oceanic parameters

In this section, the abovementioned important parameters should be presented in detail that will benefit to more accurately comprehend the oceanic turbulence. In particular, the four significant parameters, such as the rate of dissipation of kinetic energy per unit mass of fluid, the rate of dissipation of mean-squared temperature, the Kolmogorov microscale, and the ratio of temperature and salinity contributions to the refractive index spectrum, will be mainly involved in the following subsections.

2.2.1. The rate of dissipation of kinetic energy per unit mass

The rate of dissipation of the kinetic energy of the turbulent motion per unit mass of fluid through viscosity to heat is usually denoted by ε, which can be expressed as [6].

$$\varepsilon = (v/2) < s_{ij}s_{ij} > \quad (2)$$

where v is the kinematic viscosity, the tensor s_{ij} is given by $s_{ij} = \left(\partial u_i/\partial x_j + \partial u_j/\partial x_i\right)$, $(i, j = 1, 2, 3)$, the velocity is written as $\vec{u} = u_1\,\vec{i}\,+ u_2\,\vec{j}\,+ u_3\,\vec{k}$, here $\left(\vec{i}, \vec{j}, \vec{k}\right)$ is the unit vector of a Cartesian coordinate system.

Recorded $u_1 = p$, $u_2 = q$, $u_3 = u$, $x_1 = x$, $x_2 = y$ and $x_3 = z$ in this section, for isotropic turbulence, the mean-squared gradients of velocity are equal (i.e., $\partial u_1/\partial x_1 = \partial u_2/\partial x_2 = \partial u_3/\partial x_3$), and which can be written as $\partial u/\partial z$ in vertical direction so that Eq. (2) reduces to the much simpler one [6].

$$\varepsilon = (15/2)v\left\langle (\partial u/\partial z)^2 \right\rangle \tag{3}$$

2.2.2. The rate of dissipation of mean-squared temperature

The effect of turbulence on the fluid temperature field can be described as the rate of dissipation of mean-squared temperature [6],

$$\chi_T = 2\kappa_T < (\partial T'/\partial x)^2 + (\partial T'/\partial y)^2 + (\partial T'/\partial z)^2 > \tag{4}$$

where κ_T is the eddy diffusion coefficients of heat, $\partial T'/\partial x$, $\partial T'/\partial y$ and $\partial T'/\partial z$ are the mean-squared gradients of temperature in three orthogonal coordinates.

In isotropic turbulence when the mean-squared gradients of temperature are the same in all directions, so that the rate of dissipation of mean-squared temperature becomes [6].

$$\chi_T = 6\kappa_T < (\partial T'/\partial z)^2 > \tag{5}$$

It is noted that the rate of dissipation of mean-squared salinity, χ_S, is defined similarly, but is harder to determine accurately because of problems in measuring salinity changes over small distances [6].

2.2.3. Kolmogorov microscale

The turbulent flow contains eddies of various sizes, and the energy is transferred from larger eddies to smaller eddies until it is drained out by viscous dissipation. Kolmogorov's hypothesis asserts that for large Reynolds numbers (i.e., inertial subrange), the small-scale structure of turbulence is statistically steady, isotropic, and locally homogeneous, and independent of the detailed structure of the large-scale components of turbulence [12]. Kolmogorov microscale is the smallest scale in turbulent flow. At the Kolmogorov scale, viscosity dominates and the turbulent kinetic energy is dissipated into heat. The length scale of the turbulent motions at which viscous dissipation becomes important must depend on factors that provide measures of the turbulent motion and of its viscous dissipation [6]. Kolmogorov length scale [13].

$$\eta = \left(v^3/\varepsilon\right)^{1/4} \tag{6}$$

where a range of η from about 6×10^{-5} m in very turbulent regions to 0.01 m in the abyssal ocean [6].

2.2.4. The ratio of temperature and salinity contributions to the refractive index spectrum

The parameter w is the ratio of temperature and salinity contributions to the refractive index spectrum given by

$$w = \frac{\alpha(\partial T/\partial z)}{\beta(\partial S/\partial z)} \tag{7}$$

where $\alpha = 2.6 \times 10^{-4}$liter/deg, $\beta = 1.75 \times 10^{-4}$liter/gram, $\partial T/\partial z$ and $\partial S/\partial z$ are, respectively, the gradients of mean temperature and salinity between the top and bottom boundaries of domain on the vertical coordinate [8].

3. Spatial coherence radius

When one coherent optical wave propagates through a random medium, various eddies impress a spatial phase fluctuation on the wave front with an imprint of the scale size [1]. The accumulation of such fluctuations on the phase leads to a reduction in the "smoothness" of the wave front [1]. Hence, turbulent eddies further away experience a smoothness of the wave front only on the order of the transverse spatial coherence radius, which Andrews and Phillips denote by ρ_0 [1]. After a wave propagates a sufficient distance, only those turbulent eddies on the order of ρ_0 or less are effective in producing further spreading and amplitude fluctuations on the wave [1]. Except for predicting random medium-induced beam spreading through the mean irradiance, the mutual coherence function (MCF) is also used to predict the spatial coherence radius of the wave at the receiver pupil plane. Obtained from the MCF, the spatial coherence radii of plane wave and spherical wave are, respectively, deduced in this section.

3.1. Plane wave

Under Rytov approximation, the wave structure function (WSF) of a plane wave propagating through isotropic and homogeneous turbulence is defined by [1].

$$D_{\text{pl}}(\rho, L) = 8\pi^2 k^2 L \int_0^\infty [1 - J_0(\kappa\rho)]\Phi_n(\kappa)\kappa d\kappa, \tag{8}$$

where k is the optical wave number related to the wavelength λ by $k = 2\pi/\lambda$, L is the path length, κ is the magnitude of spatial wave number, ρ is the separation distance between two points on the phase front transverse to the axis of propagation and $J_0(\bullet)$ is the zero-order Bessel function.

By expanding the zero-order Bessel function in power series, the WSF is written in the form

$$D_{\text{pl}}(\rho, L) = A \sum_{n=1}^\infty \frac{(-1)^{n-1}\rho^{2n}}{(n!)^2 2^{2n}} \int_0^\infty \kappa^{2n-\frac{8}{3}}\left[1 + g\kappa^{\frac{2}{3}}\right]\left(w^2 e^{-a\kappa^{\frac{4}{3}}-b\kappa^2} + e^{-c\kappa^{\frac{4}{3}}-d\kappa^2} - 2we^{-e\kappa^{\frac{4}{3}}-f\kappa^2}\right)d\kappa, \tag{9}$$

where the power spectrum given by Eq. (1) is used and the order of summation and integration is interchanged. In addition, $a = 8.284A_T\eta^{4/3}$, $b = 12.978A_T\eta^2$, $c = 8.284A_S\eta^{4/3}$, $d = 12.978A_S\eta^2$,

$e = 8.284A_{TS}\eta^{4/3}$, $f = 12.978A_{TS}\eta^2$, $g = 2.35\eta^{2/3}$ and $A = 8\pi^2 k^2(0.388 \times 10^{-8})\varepsilon^{-1/3}\chi_T/w^2$ are taken in Eq. (9).

Based on the properties of hypergeometric function and Pochhammer symbol [14], after very tedious calculations [10], the WSF of a plane wave in certain asymptotic regimes is

$$D_{pl}(\rho, L) \approx \begin{cases} 3.063 \times 10^{-7}k^2L\varepsilon^{-1/3}\dfrac{\chi_T}{w^2}\rho^2(16.958w^2 - 44.175w + 118.923), \ (\rho \ll \eta) \\ 3.063 \times 10^{-7}k^2L\varepsilon^{-1/3}\dfrac{\chi_T}{w^2}\rho^{5/3}(1.116w^2 - 2.235w + 1.119), \quad (\rho \gg \eta) \end{cases} \quad (10)$$

The separation distance at which the modulus of the complex degree of coherence (DOC) falls to $1/e$ defines the spatial coherence radius ρ_0, that is, $D(\rho_0, L) = 2$. Based on the expressions given in Eq. (10), the plane-wave spatial coherence radius can be expressed as

$$\rho_{0pl} \approx \begin{cases} \left[3.063 \times 10^{-7}k^2L\varepsilon^{-1/3}\frac{\chi_T}{2w^2}(16.958w^2 - 44.175w + 118.923)\right]^{-1/2}, (\rho_0 \ll \eta) \\ \left[3.063 \times 10^{-7}k^2L\varepsilon^{-1/3}\frac{\chi_T}{2w^2}(1.116w^2 - 2.235w + 1.119)\right]^{-3/5}, \quad (\rho_0 \gg \eta) \end{cases} \quad (11)$$

3.2. Spherical wave

Under Rytov approximation, the WSF of a spherical wave is defined by [1].

$$D_{sp}(\rho, L) = 8\pi^2 k^2 L \int_0^1 \int_0^\infty [1 - J_0(\kappa\xi\rho)]\Phi_n(\kappa)\kappa d\kappa d\xi, \quad (12)$$

Similarly, the WSF of a spherical wave is derived in [10].

$$D_{sp}(\rho, L) \approx \begin{cases} 3.063 \times 10^{-7}k^2L\varepsilon^{-1/3}\dfrac{\chi_T}{w^2}\rho^2(5.623w^2 - 14.725w + 39.641), \ (\rho \ll \eta) \\ 3.063 \times 10^{-7}k^2L\varepsilon^{-1/3}\dfrac{\chi_T}{w^2}\rho^{5/3}(0.419w^2 - 0.838w + 0.419), \ (\rho \gg \eta) \end{cases} \quad (13)$$

and the spherical-wave spatial coherence radius as [10].

$$\rho_{0sp} \approx \begin{cases} \left[3.603 \times 10^{-7}k^2L\varepsilon^{-1/3}\frac{\chi_T}{2w^2}(5.623w^2 - 14.725w + 39.641)\right]^{-1/2}, (\rho_0 \ll \eta) \\ \left[3.603 \times 10^{-7}k^2L\varepsilon^{-1/3}\frac{\chi_T}{2w^2}(0.419w^2 - 0.838w + 0.419)\right]^{-3/5}, (\rho_0 \gg \eta) \end{cases} \quad (14)$$

3.3. Discussions

Based on the formula of Eqs. (10), (11), (13), and (14), the WSF of both a plane wave and a spherical wave can be written as

$$D(\rho, L) = \begin{cases} 2(\rho/\rho_0)^2, & (\rho \ll \eta) \\ 2(\rho/\rho_0)^{5/3}, & (\rho \gg \eta) \end{cases} \quad (15)$$

Equation (15) indicates that the spatial coherence radius is the only parameter characterizing the WSF, and under Rytov approximation, the Kolmogorov five-thirds power law of wave structure function is valid for oceanic turbulence in the inertial range if the power spectrum of oceanic turbulence proposed by Nikishov is adopted.

According to Ref. [1], under Rytov approximation, the definitions of wave structure function of a plane wave and a spherical wave are given by Eqs. (8) and (12), respectively. It is known that the expression for wave structure function depends on the mutual coherence function. Rytov approximation is limited to weak fluctuations. However, for the special cases of a plane wave and a spherical wave, it has been shown that mutual coherence function derived by strong fluctuation theories is the same as that derived by Rytov approximation [1]. Only a plane-wave and a spherical-wave case are considered in this section. Thus, the results of the wave structure function and the spatial coherence radius obtained are valid in both weak and strong fluctuations.

4. Angle-of-arrival fluctuations

Angle-of-arrival (AOA) fluctuations play an important role in a diverse range of fields including atmospheric turbulence [15, 16], free space optical communication [17], ground-based astronomical observations [18], and so on.

Angle-of-arrival fluctuations of an optical wave in the plane of the receiver aperture are associated with image dancing in the focal plane of an imaging system. Fluctuations in the AOA can be described in terms of the phase structure function [1]. In order to understand it easily, let ΔS denote the total phase shift across a collecting lens of diameter $2W_G$ and Δl the corresponding optical path difference. These quantities are related to [1]

$$k\Delta l = \Delta S, \tag{16}$$

Under the geometrical optics method, the AOA is defined by [19].

$$\beta_a = \frac{\Delta l}{2W_G} = \frac{\Delta S}{2kW_G}, \tag{17}$$

Based on the homogeneous and isotropic oceanic turbulence, the mean $\langle \beta_a \rangle = 0$ will be satisfied. The variance of AOA can be expressed as [1].

$$\langle \beta_a^2 \rangle = \frac{\langle (\Delta S)^2 \rangle}{(2kW_G)^2} = \frac{D_S(2W_G, L)}{(2kW_G)^2}, \tag{18}$$

where $D_S(\rho, L)$ is the phase structure function with the radial distance $\rho = 2W_G$.

4.1. Angle-of-arrival fluctuations of plane wave

The phase structure function associated with an unbounded plane wave is given by [1].

$$D_{S\text{-pl}}(\rho, L) = 4\pi^2 k^2 L \int_0^1 \int_0^\infty \kappa \Phi_n(\kappa)[1 - J_0(\kappa\rho)] \left[1 + \cos\left(\frac{L\kappa^2\xi}{k}\right)\right] d\kappa d\xi, \tag{19}$$

where the normalized distance variable $\xi = 1 - z/L$.

Based on Eqs. (18) and (19), the AOA fluctuations for a plane wave can be expressed as [1].

$$\langle \beta_a^2 \rangle_{\text{pl}} = \frac{D_{S\text{-pl}}(2W_G, L)}{(2kW_G)^2}, \tag{20}$$

the variance of AOA in the $\rho \gg \eta$ case is presented in this section; therefore, $\rho \gg (L/k)^{1/2}$ and the geometrical optics approximation $L\kappa^2/k \ll 1$ is satisfied.

By using $\cos\left(L\kappa^2\xi/k\right) \approx 1 - L^2\kappa^4\xi^2/2k^2$ and after very tedious calculations [20], the phase structure function of a plane wave in the inertial range as,

$$\begin{aligned}
D_{S\text{-pl}}(\rho, L) \approx \varepsilon^{-1/3}\left(\chi_T/w^2\right)\rho^{5/3}[3.063 \times 10^{-7}k^2 L(1.116w^2 - 2.235w + 1.119) \\
- L^3(1.841w^2 - 40.341w + 2077)], \qquad (\rho \gg \eta)
\end{aligned} \tag{21}$$

Substituting Eq. (21) into Eq. (20), the analytical expression of AOA fluctuations for a plane wave is

$$\begin{aligned}
\langle \beta_a^2 \rangle_{\text{pl}} \approx \varepsilon^{-1/3}\left(\chi_T/w^2\right)(2W_G)^{-1/3}[3.063 \times 10^{-7}k^2 L(1.116w^2 - 2.235w + 1.119) \\
- L^3(1.841w^2 - 40.341w + 2077)], \qquad (2W_G \gg \eta)
\end{aligned} \tag{22}$$

To clarify the physical explanation, we introduce the plane-wave spatial coherence radius $\rho_{0\text{pl}}$ (see Ref. [10] in the inertial range) in Eqs. (21) and (22), and the simplified expressions of phase structure function and AOA fluctuations for plane waves can be expressed as

$$D_{S\text{-pl}}(\rho, L) \approx C_1 \left(\frac{\rho}{\rho_{0\text{pl}}}\right)^{5/3}, \qquad (\rho \gg \eta) \tag{23}$$

$$\langle \beta_a^2 \rangle_{\text{pl}} \approx 2C_1 k^{-2}(2W_G)^{-1/3}\rho_{0\text{pl}}^{-5/3}, \qquad (2W_G \gg \eta) \tag{24}$$

where $C_1 = 2\left[1 - L^2(1.84w^2 - 40.341w + 2077)/3.063 \times 10^{-7}k^2(1.116w^2 - 2.235w + 1.119)\right]$.

4.2. Angle-of-arrival fluctuations of a spherical wave

In the case of a spherical wave, the phase structure function is defined by [1].

$$D_{S\text{-sp}}(\rho, L) = 4\pi^2 k^2 L \int_0^1 \int_0^\infty \kappa \Phi_n(\kappa)[1 - J_0(\kappa\rho)] \left\{1 + \cos\left[\frac{L\kappa^2\xi(1-\xi)}{k}\right]\right\} d\kappa d\xi, \tag{25}$$

Based on Eq. (25), the AOA fluctuations for a spherical wave can be written as [1].

$$\left\langle \beta_a^2 \right\rangle_{sp} = \frac{D_{S\text{-sp}}(2W_G, L)}{(2kW_G)^2},$$ (26)

Similarly, the phase structure function of a spherical wave is expressed as

$$D_{S\text{-sp}}(\rho, L) \approx \varepsilon^{-1/3}(\chi_T/w^2)\rho^{5/3}[3.063 \times 10^{-7}k^2L(0.4196w^2 - 2.235w + 0.419) \\ - L^3(0.184w^2 - 4.034w + 207.7)], \qquad (\rho \gg \eta)$$ (27)

and the analytical expression of AOA fluctuations for a spherical wave is

$$\left\langle \beta_a^2 \right\rangle_{sp} \approx \varepsilon^{-1/3}(\chi_T/w^2)(2W_G)^{-1/3}[3.063 \times 10^{-7}k^2L(0.4196w^2 - 2.235w + 0.419) \\ - L^3(0.184w^2 - 4.034w + 207.7)], \qquad (2W_G \gg \eta).$$ (28)

Substituting the spherical-wave spatial coherence radius ρ_{0sp} (see Ref. [10]) in Eqs. (27) and (28), phase structure function and AOA fluctuations for the spherical wave are

$$D_{S\text{-sp}}(\rho, L) \approx C_2 \left(\frac{\rho}{\rho_{0sp}}\right)^{5/3}, \qquad (\rho \gg \eta)$$ (29)

$$\left\langle \beta_a^2 \right\rangle_{sp} \approx 2C_2 k^{-2}(2W_G)^{-1/3}\rho_{0sp}^{-5/3}, \qquad (2W_G \gg \eta)$$ (30)

where $C_2 = 2\left[1 - L^2(0.184w^2 - 4.034w + 207.7)/3.063 \times 10^{-7}k^2(0.419w^2 - 0.838w + 0.419)\right]$.

4.3. Discussions

As mentioned in Section 1, the spatial coherence radius plays an important role in a heterodyne detection system. To the best of our knowledge, it is interesting to research the relation between AOA fluctuations and the spatial coherence radius, because spatial coherence radius ρ_0 represents the coherence degradation of laser beams propagating through ocean induced by the strength of turbulence. Both for plane-wave and for spherical-wave models, it is shown that AOA is inversely proportional to five-thirds order of spatial coherence radius in the inertial range. Besides, the difference of AOA fluctuations between an atmospheric turbulence and an oceanic turbulence is that the constant $C_i(i = 1, 2)$ occurs in the analytical expressions of AOA fluctuations in the latter one. Due to the constant C_i related to wavelength, propagation distance, and oceanic parameters, changes of oceanic parameter on constant C_i are investigated at fixed wavelength and propagation path [20]. To illustrate this case, $w = -1$ (i.e., $C_1 = 1.966$, $C_2 = 1.991$) is chosen in the theoretical calculations, and it is clear that constant C_i approximates to 2 both for plane and for spherical waves; thus, Eqs. (24) and (30) are consistent with the expressions of AOA fluctuations in atmospheric turbulence. In Ref. [20], changes of AOA fluctuations versus ρ_0 (i.e., ρ_{0pl} in a plane-wave case and ρ_{0sp} for a spherical-wave model, respectively) are illustrated in **Figure 1**. Both for a plane wave and for a spherical wave, AOA fluctuations decrease as ρ_0 increases. It is clear that AOA fluctuations have a

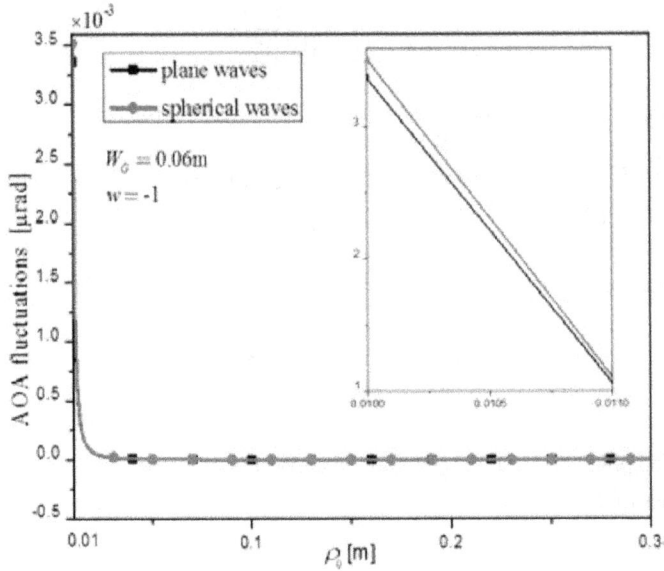

Figure 1. Changes of AOA fluctuations for oceanic turbulence versus ρ_0 in two models [20].

smaller value when the larger spatial coherence radius is obtained, and the AOA fluctuations of a spherical wave are larger than that of a plane wave at the fixed ρ_0. In terms of the influences of spatial coherence radius, the AOA fluctuations of plane and spherical waves in oceanic turbulence have the similar behavior to that of atmospheric turbulence.

5. Beam wander

Movement of the short-term beam instantaneous center (or "hot spot") is commonly called beam wander [1]. This phenomenon can be characterized statistically by the variance of the hot spot displacement along an axis or by the variance of the magnitude of the hot spot displacement [1]. An estimate of the short-term beam radius is obtained by removing beam wander effects from the long-term beam radius [1]. It is much convenient to use the geometrical optics approximation method in the turbulent area. Beam wander is an important characteristic of laser beams, which determines their utility for practical applications, such as ground-to-satellite laser communication [2, 3] and global quantum communication [4].

5.1. A general model

The far-field angular spread of a free-space propagating beam of diameter $2W_0$ is of order $\lambda/2W_0$. In the presence of optical turbulence, a finite optical beam will experience random deflections as it propagates, causing further spreading of the beam by large-scale inhomogeneities of the turbulence [1]. Over short time periods, the beam profile at the receiver moves off the bore sight and can become highly skewed from Gaussian so that the instantaneous center of the beam is randomly displaced [1]. According to Ref. [1], $W^2 T_{LS}$ describes the beam wander or the variance of the instantaneous center of the beam in the receiver plane ($z = L$).

Based on the introduction of a general model [1], beam wander can be expressed as

$$\langle r_c^2 \rangle = W^2 T_{LS}$$

$$= 4\pi^2 k^2 W^2 \int_0^L \int_0^\infty \kappa \Phi_n(\kappa) H_{LS}(\kappa, z) \left[1 - \exp\left(-\Lambda L \kappa^2 \xi^2 / k\right) \right] d\kappa dz, \tag{31}$$

where bracket < > denotes an ensemble average, W is the beam radius in the free space at receiver, Λ represents Fresnel ration of beam at receiver, and $H_{LS}(\kappa, z)$ is the large-scale filter function, respectively.

The large-scale filter function is [1].

$$H_{LS}(\kappa, \xi) = \exp\left\{ -\kappa^2 W_0^2 \left[\left(\Theta_0 + \overline{\Theta}_0 \xi\right)^2 + \Lambda_0^2 (1 - \xi)^2 \right] \right\}, \tag{32}$$

where $\Theta_0 = 1 - \overline{\Theta}_0$. W_0, Θ_0 and Λ_0 are the beam radius, the beam curvature parameter, and Fresnel ration of beam at transmitter, respectively [1].

Because beam wander is caused mostly by a large-scale turbulence near the transmitter, the last term can be dropped in Eq. (32) and the geometrical optics approximation is [1].

$$1 - \exp\left(-\Lambda L \kappa^2 \xi^2 / k\right) = \Lambda L \kappa^2 \xi^2 / k, \qquad L \kappa^2 / k \ll 1. \tag{33}$$

Substituting from Eqs. (1), (32), and (33) into Eq. (31), (31) leads to

$$\langle r_c^2 \rangle = 0.388 \times 10^{-8} \times 4\pi^2 k W^2 L \Lambda \varepsilon^{-1/3} \left(\chi_T / w^2\right)$$

$$\times \int_0^1 \int_0^\infty \kappa^{-2/3} \left[1 + 2.35(\kappa\eta)^{2/3} \right] \left[w^2 \exp\left(-A_T \delta\right) + \exp\left(-A_S \delta\right) - 2w \exp\left(-A_{TS}\delta\right) \right] \tag{34}$$

$$\exp\left[-\kappa^2 W_0^2 \left(\Theta_0 + \overline{\Theta}_0 \xi\right)^2 \right] \xi^2 d\kappa d\xi$$

Equation (34) is applicable for collimated, divergent, or focused Gaussian-beam waves, and it can represent our general expression for the variance of beam wander displacement under weak irradiance fluctuations.

5.2. Special cases

In this section, two special cases (i.e., collimated beam and focused beam) are analyzed.

For collimated beam ($\Theta_0 = 1$), the beam wander can be simplified as

$$\langle r_c^2 \rangle_{coll} = 0.517 \times 10^{-8} \times \pi^2 k W^2 L \Lambda \varepsilon^{-1/3} \left(\chi_T / w^2 \right) \left(\alpha w^2 - 2\beta w + \gamma \right), \tag{35}$$

where

$$\begin{aligned}
\alpha &= 37.9244 \eta^4 A_T^2 \left(W_0^2 + 12.978 A_T \eta^2 \right)^{-11/6} + 15.2043 \eta^{8/3} A_T^2 \left(W_0^2 + 12.978 A_T \eta^2 \right)^{-3/2} \\
&\quad - 9.03014 \eta^{8/3} A_T \left(W_0^2 + 12.978 A_T \eta^2 \right)^{-7/6} - 4.67544 \eta^{4/3} A_T \left(W_0^2 + 12.978 A_T \eta^2 \right)^{-5/6} \\
&\quad + 2.08263 \eta^{4/3} \left(W_0^2 + 12.978 A_T \eta^2 \right)^{-1/2} + 2.78316 \left(W_0^2 + 12.978 A_T \eta^2 \right)^{-1/6}.
\end{aligned} \tag{36}$$

and β, γ can be obtained from Eq. (36) if A_T is replaced by A_{TS}, A_S, respectively.

For focused beam ($\Theta_0 = 0$), the beam wander is expressed as

$$\langle r_c^2 \rangle_{focu} = 0.517 \times 10^{-8} \times \pi^2 k W^2 L \Lambda \varepsilon^{-1/3} \left(\chi_T / w^2 \right) \left(\alpha' w^2 - 2\beta' w + \gamma' \right), \tag{37}$$

where

$$\begin{aligned}
\alpha' &= \int_0^1 \Bigl[37.9244 \eta^4 A_T^2 \left(W_0^2 \xi^2 + 12.978 A_T \eta^2 \right)^{-11/6} + 15.2043 \eta^{8/3} A_T^2 \left(W_0^2 \xi^2 + 12.978 A_T \eta^2 \right)^{-3/2} \\
&\quad - 9.03014 \eta^{8/3} A_T \left(W_0^2 \xi^2 + 12.978 A_T \eta^2 \right)^{-7/6} - 4.67544 \eta^{4/3} A_T \left(W_0^2 \xi^2 + 12.978 A_T \eta^2 \right)^{-5/6} \\
&\quad + 2.08263 \eta^{4/3} \left(W_0^2 \xi^2 + 12.978 A_T \eta^2 \right)^{-1/2} + 2.78316 \left(W_0^2 \xi^2 + 12.978 A_T \eta^2 \right)^{-1/6} \Bigr] \xi^2 d\xi.
\end{aligned} \tag{38}$$

and β', γ' can be obtained from Eq. (38) when A_T is replaced by A_{TS}, A_S, respectively.

To atmospheric turbulence, the focused beam case leads to a greater beam wander variance for the same size beam at the transmitter as that for the collimated beam [1]. However, because of the complexity of oceanic power spectrum, the analytical expressions of collimated beam and focused beam are also less concise than those of atmospheric turbulence. Therefore, it is not simple to distinguish whose variance of beam wander is the larger one directly. In Section 5.4, numerical calculations are used to discuss the abovementioned property of different beams.

5.3. Dimensionless quantity B_W

In order to obtain influences of beam wander on laser beam propagation through oceanic turbulence, the relation between beam wander and the turbulence-induced beam spot size is investigated in detail by using theoretical and numerical methods. Using the dimensionless quantity $B_W = \langle r_c^2 \rangle / W^2 (1 + T)$, where $T = 4\pi^2 k^2 L \int_0^1 \int_0^\infty \kappa \Phi_n(\kappa) \left(1 - e^{-\Lambda L \kappa^2 \xi^2 / k} \right) d\kappa d\xi$ [1]. Based on Ref. [21], the quantity in oceanic turbulence can be expressed as $T = 0.517 \times 10^{-8} \pi^2 k L^2 \Lambda \varepsilon^{-1/3} \chi_T \left(67.832 w^2 - 176.699 w + 475.692 \right) / w^2$. For the dimensionless quantity, B_W is more informative than merely $\langle r_c^2 \rangle$ about the practical significance of the beam wander. The quantity B_W can be used to investigate the proportion of beam wander $\langle r_c^2 \rangle$ to the turbulence-induced beam spot size $W^2 (1 + T)$ [1]. In **Figure 2**, the dimensionless quantity B_W as a function of the three oceanic parameters is plotted. It is shown that the larger value of B_W is related to smaller

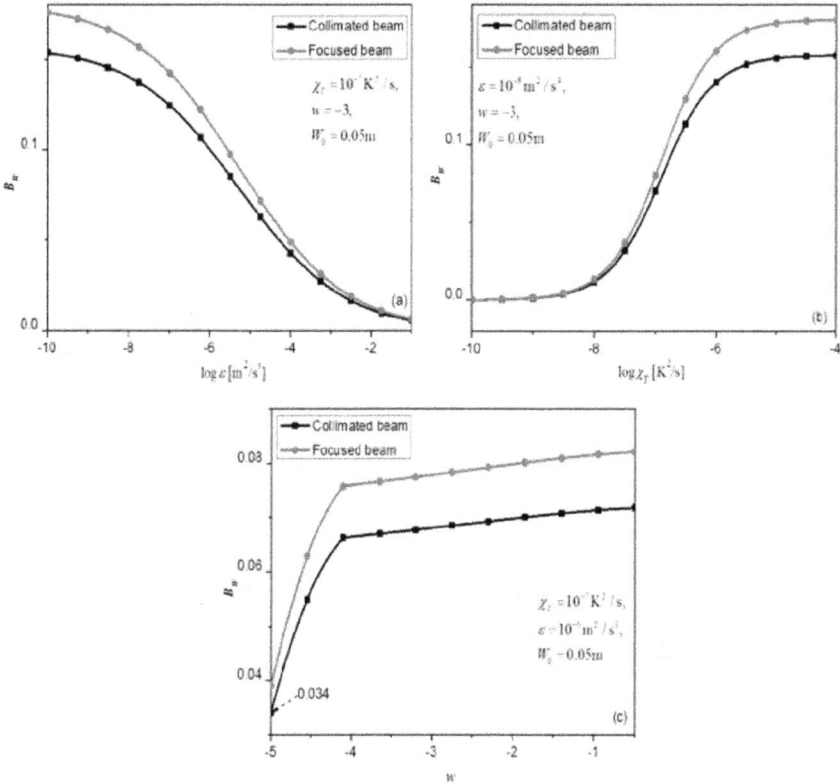

Figure 2. Dimensionless quantity B_W for collimated and focused beam versus (a) $\log \varepsilon$, (b) $\log \chi_T$ and (c) w [22].

$\log \varepsilon$, larger $\log \chi_T$, and salinity-induced dominant. The beam wander of collimated beam has less influence on turbulence-induced beam spot size compared to that of focused beam. Furthermore, beam wander plays an unignorable role in turbulence-induced beam spot size because all the values of B_W are larger than 0.034 (or 3.4%) shown in **Figure 2**. The beam wander effect cannot be ignored on laser beam propagating through ocean.

5.4. Definition of relative beam wander

To obtain the difference of beam wander among various beam types, the relative beam wander $\langle r_c^2 \rangle_R$ based on focused and collimated beam is defined, which can be expressed as

$$\langle r_c^2 \rangle_R = \langle r_c^2 \rangle_{focu} - \langle r_c^2 \rangle_{coll}$$
$$= 0.517 \times 10^{-8} \times \pi^2 k W^2 L \Lambda \varepsilon^{-1/3} (\chi_T/w^2) \left[(\alpha' - \alpha)w^2 - 2(\beta' - \beta)w + (\gamma' - \gamma) \right]. \tag{39}$$

It is clear that the increment $\langle r_c^2 \rangle_R' = \langle r_c^2 \rangle_{focu} - \langle r_c^2 \rangle_{arbi}$ or $\langle r_c^2 \rangle_R' = \langle r_c^2 \rangle_{arbi} - \langle r_c^2 \rangle_{coll}$ of arbitrary beam type (i.e., $0 < \Theta_0 < 1$) is smaller than that of Eq. (39). Choosing the same turbulent strength and beam radius (i.e., $\varepsilon = 10^{-5} \text{m}^2/\text{s}^3$, $\chi_T = 10^{-7} \text{K}^2/\text{s}$, $w = -3$ and $W_0 = 0.05\text{m}$) shown in **Figure 3**, the value of $\langle r_c^2 \rangle_R / \langle r_c^2 \rangle_{focu}$ is equal to 0.124. For arbitrary beam, it is clear that the ratio of $\langle r_c^2 \rangle_R / \langle r_c^2 \rangle_{focu}$ is smaller than 0.124. The extension to arbitrary oceanic parameters and beam radius is straightforward.

From Section 5.3, it is impossible to avoid the beam wander effect for laser propagation; therefore, achieving small beam wander is imperative. In this section, the relative beam wander describes the increment of beam wander between focused and collimated beam, and this quantity benefits us to select how to obtain small value of beam wander. Based on Eq. (39), it is feasible to obtain small beam wander as long as the arbitrary beam type is under $\Theta_0 \rightarrow 1$ condition. **Figure 3** shows that the larger beam radius leads to a smaller value of beam wander. In the practical applications, it is an effective feasible solution to achieve a small beam wander effect when beam curvature parameter at transmitter Θ_0 is approximate to 1 and beam radius is appropriately large. It is also meaningful to select favorable beam parameters which are less sensitive to turbulence in laser propagation applications.

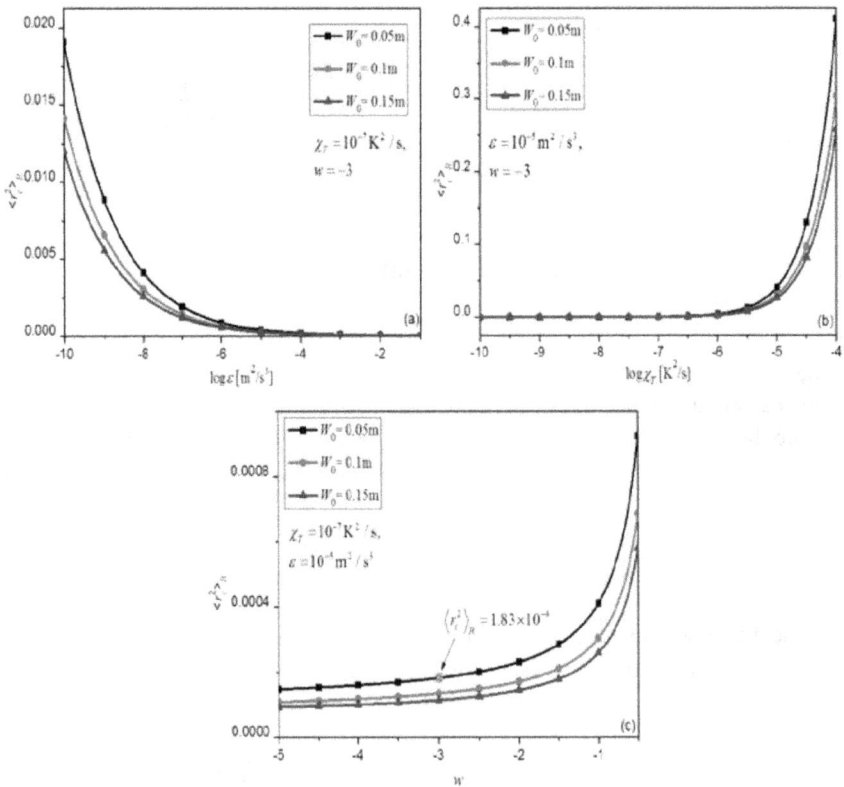

Figure 3. Relative beam wander with various beam radii versus (a) $\log \varepsilon$, (b) $\log \chi_T$ and (c) w [22].

6. Conclusions

In summary, this chapter has used the same idea to convert atmospheric turbulence concerning beam wave propagation to corresponding oceanic turbulence. The general characteristics of an optical wave propagating through the ocean are greatly affected by small fluctuations in the refractive index that are the direct consequence of small temperature and salinity fluctuations transported by the turbulent motion of the ocean. Therefore, the propagation process of one optical wave suffers beam spreading, loss of spatial coherence, angle-of-arrival fluctuations, beam wander, and so on. In this chapter, three important statistical quantities including spatial coherence radius, angle-of-arrival fluctuations, and beam wander have been investigated.

The analytical formulae for the wave structure function and the spatial coherence radius of a plane wave and a spherical wave propagating through oceanic turbulence have been derived in Section 3, which are valid in both weak and strong fluctuations. It has been shown that under Rytov approximation, the Kolmogorov five-thirds power law of wave structure function is also valid for the oceanic turbulence in the inertial range if the power spectrum of oceanic turbulence proposed by Nikishov is adopted. These results are of considerable theoretical and practical interest for operations in communication, imaging, and sensing systems involving turbulent underwater channels.

Furthermore, spatial coherence radius can be described as the coherence degradation of laser beams propagating through ocean induced by the strength of turbulence; thus, the relation between angle-of-arrival fluctuations and the spatial coherence radius has been researched. Both for a plane wave and for a spherical wave, it is shown that the angle-of-arrival fluctuations are inversely proportional to five-thirds order of spatial coherence radius in the inertial range. In terms of the influences of spatial coherence radius, the angle-of-arrival fluctuations of plane and spherical waves in oceanic turbulence have the similar behavior to that of atmospheric turbulence.

In addition, based on the oceanic power spectrum, the beam wander effect has been studied with analytical and numerical methods in weak fluctuation theory, and the analytical expressions for beam wander of collimated and focused beam in oceanic turbulence have also been derived. For the dimensionless quantity, B_W, the relation between beam wander and the turbulence-induced beam spot size has been investigated. It is shown that the beam wander of collimated beam has less influence on turbulence-induced beam spot size compared to that of focused beam. Particularly, according to the definition of the relative beam wander, it is shown that the relative beam wander is small when the value of beam curvature parameter at transmitter Θ_0 is close to 1 (i.e., $\Theta_0 \rightarrow 1$) and beam radius W_0 is properly large.

In this chapter, the classical treatments of optical wave propagation have been concerned with part of special cases, such as uniform plane wave, spherical wave, collimated beam, and focused beam. The results presented serve as a foundation for the study of optical beam propagation in oceanic turbulence, which may provide an essential support for further researches in applications for underwater communicating, imaging, and sensing systems. Thus, these simple optical wave models are useful in describing certain aspects of wave propagation in oceanic turbulence. However, due to inherent infinite extent, these models are not adequate in describing laser beams when finite size of the transmitted wave and high-order Gaussian wave should be taken into account in the near future.

Acknowledgements

Lu Lu and Zhiqiang Wang wrote this manuscript. Lu Lu, Zhiqiang Wang, and Chengyu Fan arranged the structure of this manuscript. Lu Lu and Xiaoling Ji performed theoretical calculations and physical analysis in Section 2. Lu Lu, Zhiqiang Wang, Pengfei Zhang, Chunhong Qiao, Jinghui Zhang, and Chengyu Fan performed Sections 3–5. Lu Lu deduced the analytical expressions. Zhiqiang Wang and Pengfei Zhang performed the numerical simulations and analyzed data. Pengfei Zhang, Jinghui Zhang, Chunhong Qiao, and Chengyu Fan revised this manuscript. All authors contributed this work equally and acknowledged the support by the National Natural Science Foundation of China (NSFC) under grants 61405205 and 61475105.

Conflict of interest

The authors declare no conflict of interest.

Notes

All authors are very much thankful to the valuable suggestions in the spatial coherence radius section by Prof. Yahya Baykal.

Author details

Zhiqiang Wang [1,2], Lu Lu [3*], Pengfei Zhang [1], Chunhong Qiao [1*], Jinghui Zhang [1], Chengyu Fan [1*] and Xiaoling Ji [4]

*Address all correspondence to: lulu19900101@berkeley.edu; chqiao@aiofm.ac.cn; cyfan@aiofm.ac.cn

1 Key Laboratory of Atmospheric Optics, Anhui Institute of Optics and Fine Mechanics, Chinese Academy of Sciences, Hefei, China

2 University of Science and Technology of China, Hefei, China

3 University of California, Berkeley, CA, USA

4 Department of Physics, Sichuan Normal University, Chengdu, China

References

[1] Andrews LC, Phillips RL. Laser Beam Propagation Through Random Media. 2nd ed. Bellingham: SPIE; 2005

[2] Dios F, Rubio JA, Rodríguez A, Comerón A. Scintillation and beam-wander analysis in an optical ground station-satellite uplink. Applied Optics. 2004;**43**:3866-3873. DOI: 10.1364/AO.43.003866

[3] Guo H, Luo B, Ren Y, Zhao S, Dang A. Influence of beam wander on uplink of ground-to-satellite laser communication and optimization for transmitter beam radius. Optics Letters. 2010;**35**:1977-1979. DOI: 10.1364/OL.35.001977

[4] Vasylyev DY, Semenov AA, Vogel W. Toward global quantum communication: Beam wandering preserves nonclassicality. Physical Review Letters. 2012;**108**:220501. DOI: 10.1103/PhysRevLett.108.220501

[5] Rothschild BJ, editor. Toward a Theory on Biological-Physical Interactions in the World Ocean. NATO ASI Series (Series C: Mathematical and Physical Sciences). Vol. 239. Dordrecht: Springer; 1988. pp. 215-234. DOI: 10.1007/978-94-009-3023-0_1

[6] Thorpe SA. An Introduction to Ocean Turbulence. Cambridge: Cambridge University Press; 2007

[7] Ferrari R, McWilliams JC, Canuto VM, Dubovikov M. Parameterization of Eddy fluxes near oceanic boundaries. Journal of Climate. 2008;**21**:2770-2789. DOI: 10.1175/2007JCLI1510.1

[8] Nikishov VV, Nikishov VI. Spectrum of turbulent fluctuations of the sea-water refraction index. International Journal of Fluid Mechanics Research. 2000;**27**:82-98. DOI: 10.1615/InterJFluidMechRes.v27.i1.70

[9] Lu W, Liu L, Sun J. Influence of temperature and salinity fluctuations on propagation of partially coherent beams in oceanic turbulence. Journal of Optics A: Pure and Applied Optics. 2006;**8**:1052-1058. DOI: 10.1088/1464-4258/8/12/004

[10] Lu L, Ji XL, Baykal Y. Wave structure function and spatial coherence radius of plane and spherical waves propagating through oceanic turbulence. Optics Express. 2014;**22**(22): 27112-27122. DOI: 10.1364/OE.22.027112

[11] Farwell N, Korotkova O. Intensity and coherence properties of light in oceanic turbulence. Optics Communication. 2012;**285**(6):872-875. DOI: 10.1016/j.optcom.2011.10.020

[12] Pao YH. Structure of turbulent velocity and scalar fields at large wavenumbers. Physics of Fluids. 1965;**8**:1063-1075. DOI: 10.1063/1.1761356

[13] Landahl MT, Mollo-Christensen E. Turbulence and Random Processes in Fluid Mechanics. 2nd ed. Cambridge: Cambridge University Press; 1992. p. 10

[14] Andrews LC. Special Functions of Mathematics for Engineers. 3rd ed. Oxford: SPIE and Oxford University; 1998

[15] Hill RJ. Review of optical scintillation methods of measuring the refractive index spectrum, inner scale and surface fluxes. Waves in Random and Complex Media. 1992;**2**:179-201. DOI: 10.1088/0959-7174/2/3/001

[16] Vilcheck MJ, Reed AE, Burris HR, Scharpf WJ, Moore CI, Suite MR. Multiple methods for measuring atmospheric turbulence. Proceedings of SPIE. 2002;**4821**:300-309. DOI: 10.1117/12.450631

[17] Eyyuboglu HT, Baykal Y. Analysis of laser multimode content on the angle-of-arrival fluctuations in free-space optics access systems. Optical Engineering. 2005;**44**:056002. DOI: 10.1117/1.1905383

[18] Masciadri E, Vernin J, Bougeault P. 3D mapping of optical turbulence using an atmospheric numerical model. I. A useful tool for the ground-based astronomy. Astronomy and Astrophysics Supplement Series. 1999;**137**(1):185-202. DOI: 10.1051/aas:1999475

[19] Tatarskii VI. Wave propagation in a Turbulence Medium, trans. In: Silverman RA, editor. New York: McGraw-Hill; 1961

[20] Lu L, Wang ZQ, Zhang PF, Qiao CH, Fan CY, Zhang JH, Ji XL. Phase structure function and AOA fluctuations of plane and spherical waves propagating through oceanic turbulence. Journal of Optics. 2015;**17**:085610. DOI: 10.1088/2040-8978/17/8/085610

[21] Lu L, Zhang PF, Fan CY, Qiao CH. Influence of oceanic turbulence on propagation of a radial Gaussian beam array. Optics Express. 2015;**23**:2827-2836. DOI: 10.1364/OE.23.002827

[22] Lu L, Wang ZQ, Zhang PF, Zhang JH, Fan CY, Ji XL, Qiao CH. Beam wander of laser beam propagating through oceanic turbulence. Optik. 2016;**127**:5341-5346. DOI: 10.1016/j.ijleo.2016.01.190

Chapter 4

Generation, Evolution, and Characterization of Turbulence Coherent Structures

Zambri Harun and Eslam Reda Lotfy

Additional information is available at the end of the chapter

http://dx.doi.org/10.5772/intechopen.76854

Abstract

Turbulence stands as one of the most complicated and attractive physical phenomena. The accumulated knowledge has shown turbulent flow to be composed of islands of vortices and uniform-momentum regions, which are coherent in both time and space. Research has been concentrated on these structures, their generation, evolution, and interaction with the mean flow. Different theories and conceptual models were proposed with the aim of controlling the boundary layer flow and improving numerical simulations. Here, we review the different classes of turbulence coherent structures and the presumable generation mechanisms for each. The conceptual models describing the generation of turbulence coherent structures are generally classified under two categories, namely, the bottom-up mechanisms and the top-down mechanisms. The first assumes turbulence to be generated near the surface by some sort of instabilities, whereas the second assigns an active role to the large outer layer structures, perhaps the turbulent bulges. Both categories of models coexist in the flow with the first dominating turbulence generation at low Reynolds number and the second at high Reynolds number, such as the case in the atmospheric boundary layer.

Keywords: boundary layer, turbulence, coherent structures, generation, ejection, sweep

1. Introduction

Turbulent flow is the most common flow in industrial applications and atmospheric phenomena. The random motion inherent in the flow contributes the largest share of fluid mixing and interaction with solid surfaces (e.g. friction, heating, and pollutant dispersion). Natural ventilation in modern cities, flow-induced vibrations of large civil structures, performance of windmills, etc. keep the topic ever interesting. Hundreds of researches have been devoted to the subject aiming at characterizing this randomness. The main objectives are (1) to set an exact solution (or at least a mathematical model) to the turbulent flow problem and (2) to control it, for example, modify vortices and thereby reduce the drag on surfaces.

The random appearance of turbulent flow is violated by many well-established evidences. For instance, compared to a random signal, the turbulent velocity signal displays non-zero trends in both the energy spectrum (**Figure 1**) and autocorrelation (**Figure 2**) analyses. These examples, among many others, reveal the existence of organized motions within the irregular background. These organized motions are termed turbulence coherent structures (TCSs). Thus, TCSs are either vortices or uniform-momentum regions within the turbulent flow; these structures maintain their coherence over remarkable extents in time and space. The TCSs play a prominent role in the transport and mixing processes within the turbulent flow.

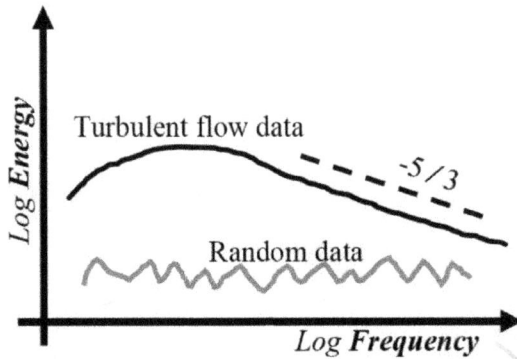

Figure 1. Energy spectrum of turbulent flow compared to a random signal.

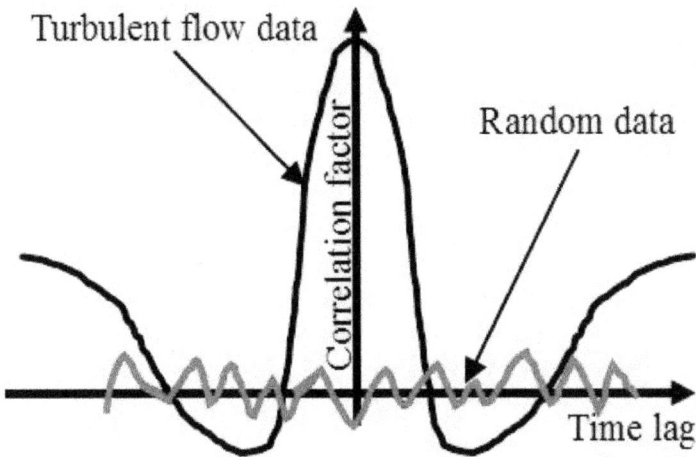

Figure 2. Autocorrelation function of turbulent flow compared to a random signal.

Accordingly, few friction-reduction schemes target manipulating the TCSs. Furthermore, few researches attribute large parts of the loading on windmills to the TCSs. It follows that the understanding of TCSs is inevitable in solving and controlling turbulent flows.

In this chapter, we focus on the basic kinds of TCSs and their presumable generation mechanisms. We start with the *hairpin vortex*, which is the elementary building block of TCSs. We characterize the hairpin vortex and detail its popular *bursting* theory of generation. Afterwards, we discuss the *vortex packets* and *superstructures*, which form the turbulent/non-turbulent interface bulges and contribute around half the turbulent kinetic energy and turbulent transport. Finally, we review the theories of TCS generation in high-Reynolds number flows.

2. The hairpin vortex

The first conceptual model for TCSs was proposed by Theodorsen [1]. From his observations, he noticed the turbulence vortex to take a *hairpin* or *horseshoe* shape with the legs aligned streamwise and the head located downstream and curved up, **Figure 3**. Theodorsen applied a vorticity-based version of Navier-Stokes equations to the proposed model vortex. He hypothesized the head to be inclined at an angle 45° to the mean flow direction since it subjects the hairpin to the maximum stretching from the mean flow and hence achieves the maximum turbulence production. The legs (streamwise vortices) induce upward flow on the head, which causes it to be lifted up. The vortex is then subjected to stretching by the mean flow since the head lies in a higher-velocity region than the legs (shear effect), see **Figure 4**. This shear causes the vortex to extend in length and compress in diameter. Consequently, the vorticity intensifies, that is, rotation becomes faster and hence more lifting force is generated and the head moves up further. This sequence is resisted only by the shear stress which, although lengthening the vortex, exerts a restoring moment on the head to return it to the zero-shear horizontal position. The inclination angle of the vortex will depend on the balance between the two conflicting effects. The hairpin vortex model was first verified experimentally by Head and Bandyopadhyay [2] through a smoke visualization experiment. They measured the

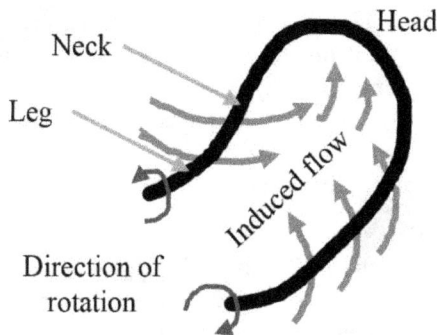

Figure 3. Illustration of the hairpin vortex model; vortex head lifting by induction from the legs.

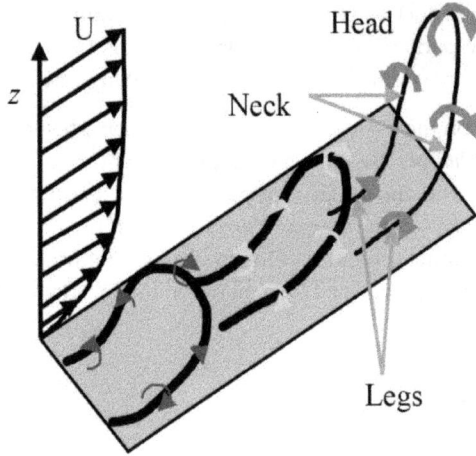

Figure 4. Hairpin vortex stretching.

angle and found it to fall well between 40 and 50°. As justified by Head and Bandyopadhyay, the angle of principal stress in the pure shear flow is 45°. Brief reviews of Theodorsen's paper are found in [2–4].

Long before the first documented visualization of the hairpin vortices, Theodorsen's model was confirmed, partly, by the correlation analysis of Townsend [5] and Grant [6] which was developed in [7]. Townsend depicted the dominant TCSs as randomly located couples of counter-rotating vortices aligned in the streamwise direction. These vortices (eddies) are of cone-like structure with the vertex upstream and the base downstream. An eddy size scales with its distance from the wall; hence his model is named the *attached eddy* model. The attached eddies can be thought of as headless hairpin vortices. In addition to Townsend, Willmarth, and Tu [8] conducted pressure-velocity correlation analysis that demonstrated the turbulence coherent structure as a transverse row of inclined triangular tubular vortices.

The symmetric structure of the hairpin vortex is the exception rather than the rule [3, 9–13]. The turbulence-inherent perturbations of the background flow cause the generated hairpin to be born distorted, for example, one-legged. Zhou et al. [14] examined the conditions to synthesize hairpin vortices by utilizing direct numerical simulation (DNS). They found the asymmetric hairpins to form more readily in rapid succession and at smaller streamwise separation. In the same article, Zhou et al. explained how the induction of the hairpin legs could cause the head to deform into an Ω-shaped structure. Thus, the transverse vortex can exist in many forms; cane, hairpin, horseshoe, or Ω -shaped vortices and deformed versions.

With or without the legs being attached to the wall, a hairpin vortex persistently rises across the boundary layer. The vortex envelope expands in the wall-normal and spanwise directions. The vortex core enlarges and weakens due to shear relaxation in the outer layer. Along the way up, hairpin-hairpin merging occurs to form larger and stronger vortices [15]. The hairpins align streamwise in groups (packets) to form the bulges at the edge of the turbulent boundary layer [2, 16–18], see **Figure 5**. Ultimately, under excessive stretching, the hairpin legs get very close

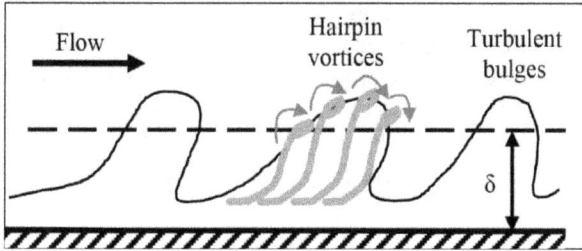

Figure 5. Hairpin vortices may reach the end of the TBL to compose the turbulent/non-turbulent interface bulges.

and cancel each other i.e. the vortex dies [18]. However, the vortex dies anyway after a while by viscous diffusion. The debris from the dead eddies is convected away from the wall and undergoes stretching and distortion by live eddies to form isotropic fine-scale eddies surrounding the attached eddies [15]. Lozano-Durán and Jiménez [19] performed a DNS to inspect the evolution of coherent structures. They argue the tendency of eddies to remain small and die shortly; few eddies only attach to the wall and expand self-similarly across the logarithmic layer. These hold-on for lifetimes, which are proportional to their distances from the wall. These eddies are responsible for the vast majority of momentum transport. The hairpin vortices transport the low-momentum fluid from the wall layer to the outer layer. The hairpins are the main elements responsible also for vortex regeneration and hence the self-sustenance of flow turbulence [20, 21].

3. From eddies to turbulence

The coordinate system is defined by x, y, and z as the streamwise, spanwise, and wall-normal directions and the velocity components are given by u, v, and w, respectively. The time-mean and fluctuating components are referred to by capital letters and (') signs. The stirring effect of the vortices is illustrated in **Figure 6**. The rotation of the hairpin vortices, either the head or legs, disturbs the fluid in two ways. The low-speed fluid ($-u'$) from the bottom layers is pumped upward ($+w'$), an event named *second-quadrant* or Q2 event, whereas the high-speed fluid ($+u'$) from the top layers is pumped downward ($-w'$), an event named *fourth-quadrant* or Q4 event. Experiments held by [22] have proven the Reynolds' turbulent stresses $\overline{u'w'}$ to be formed up of

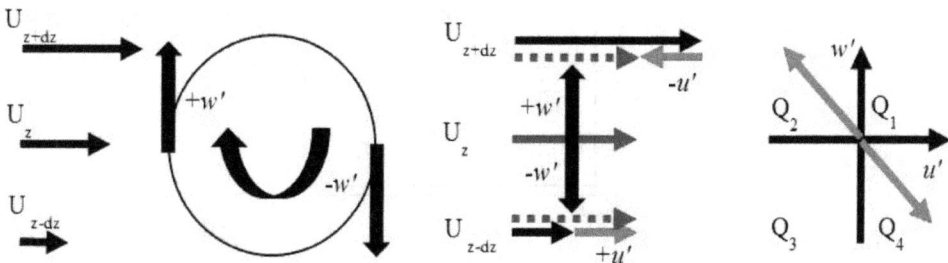

Figure 6. Q2 and Q4 events.

mainly Q2 and Q4 events. That means, the Q2 and Q4 fluctuations are more probable than Q1 and Q3 ones [3]. The turbulent kinetic energy is defined as $k = \frac{1}{2}(\overline{u'^2 + v'^2 + w'^2})$. The literature, especially of numerical analysis, defines the production of the turbulent kinetic energy as $\overline{u'w'}\frac{\partial U}{\partial z}$ [23]. Hence, it can be said that hairpin vortices by their stirring action are the *turbulence producers* i.e. they cause the fluctuations read by the hotwire probe or pressure transducer. Some researchers like to make a shortcut by identifying vortex generation as *turbulence production*.

4. Generation of the hairpin vortex

The generation of hairpin vortices is attributed *mainly* to what is called the *bursting* process [3, 10], which occurs in the buffer layer. Before proceeding with the bursting process, it is better to introduce the *low-speed streak*, which is the key element in the bursting process. The low-speed streaks are long, narrow, uniform-momentum regions aligned quasi-streamwise, see **Figure 7**. They exist exclusively in the inner layer (below z^+~10) and move downstream at speeds lower than the mean flow speed (where $z^+ = \frac{u^* z}{v}$, v is the kinematic viscosity, and u^* is the friction velocity). The streaks were first observed by Francis Hama [24] and concurrently by Ferrell et al. [25] in tube flow by injecting dye through a slot in the wall in the first experiment and by flushing a flow of colored water by a clear fluid in the second experiment.

The streaks can extend in length to 1000 viscous (wall) units [26, 27] (one viscous unit $=\frac{u^*}{v}$) and in width to 20 viscous units [28]. The transverse spacing between streaks depends on the turbulent Reynolds number, $Re_\tau = \frac{\delta u^*}{v}$, or momentum-thickness Reynolds numbers, $Re_\theta = \frac{\theta U_\infty}{v}$, [29] (where θ is the momentum thickness). The streak transverse spacing is equal to $\lambda_y^+ = 100$ at $Re_\theta = 2000$ [30] and widens to $\lambda_y^+ = 200$ at $Re_\theta = 40{,}000$ [8, 31]. Nevertheless, a later study by Smith and Metzler [32] for $740 < Re_\theta < 5830$ suggested the low-speed streaks to have an invariant spacing of $\lambda_y^+ = 100$. This value was found to provide the maximum energy amplification of a perturbation in turbulent flow [33–35]. It is worth noting that a flow domain of spanwise extent less than $\lambda_y^+ = 100$ cannot sustain turbulence [36].

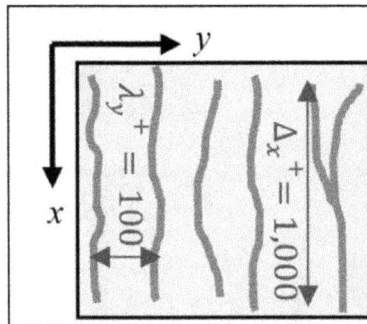

Figure 7. Low-speed streaks.

5. The bursting process

The bursting process as described by Kline et al. [37] and Kim et al. [29], with updates from later observations, passes through three stages:

1. *Streak-lifting*: The streak moves downstream and migrates gradually from the wall. The streak becomes thinner as it drifts outward. After a certain critical distance, the streak is lifted up rapidly away from the wall. This streak transports the low-momentum fluid near the wall to the upper layers, which causes an inflection in the streamwise instantaneous velocity profile. A spanwise *shear layer* is formed atop the streak upstream or downstream the crest. The shear layer (*vorticity layer*) is a circulating area of fluid of elliptic or generally non-circular shape. The shear layer rolls up in a circular form to generate a spanwise vortex (circular shear layer). The vortex is then stretched and lifted by the mean shear to form a streamwise and/or a hairpin vortex. Thus, streamwise and/or spanwise vortices propagate downstream the inflection point.

2. *Oscillation*: When the streak reaches a height of $z+ = 8 - 12$, it starts to oscillate. The oscillations are three-dimensional, that is, can be seen in both $x - z$ and $x - y$ planes and tend to be regular and organized.

3. *Break-up*: After a certain number of oscillations (3–10) the motion turns to be random and violent. This ends up with the streak broken-up and disappeared.

It follows then a quiescent period before the cycle is repeated. An illustration of the bursting process is shown in **Figure 8**. The oscillations of the streak are actually due to the formation and stretching of the born vortices. The concluding violent motion is imputed to the vortex stretching under the combined effect of turbulent background and successive-ascent through higher-faster-layers [38]. Smith and Metzler [32] discovered that the streak does not break down after the bursting process. It rather persists owing to the reinforcement by the legs of the new hairpins.

Corino and Brodkey [39] complemented the picture with a *sweep* at the onset of the burst sequence and multiple *ejections* of low-momentum fluid followed by a *sweep* at the end of process. An ejection is a Q2 event, while a sweep is a Q4 event. The sweep at the onset of the burst may be responsible for lifting the streak. As elucidated by Grass [40], the sweep (inrush) stream triggers the bursting process, while the ejection stream is a consequence of the bursting process and can extend across the entire boundary layer. On the other hand, Nakagawa and Nezu [41] and Smith [42] suggested the final ejections and sweep to be invoked by the generated hairpin vortices. The inward side of the vortex entrains low-momentum fluid from the streak and pumps it upward. The vortex navigation over the streak appears like multiple rapid ejections, whereas the outboard side entrains high-momentum fluid from the upper layers and pumps it toward the lateral extremes of the streak. Since the vortex is already inclined to the flow direction, the ejection and sweep appear as Q2 and Q4 events, see **Figure 9**.

Smith [42] considered each burst to be responsible for generating 2–5 vortices. Kline et al. [37] also estimated the frequency of bursts and found it to match with the dominant frequency

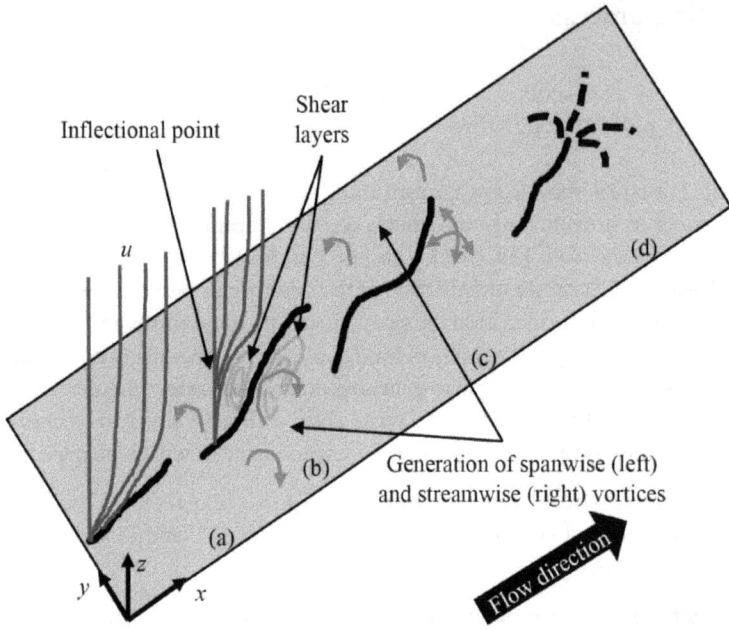

Figure 8. The bursting process as described by Kim et al. [29]. (a) Low-speed streak moving downstream and gradually lifting away from the wall. (b) Streak-lifting: the streak is lifted rapidly. (c) Oscillation: the streak starts an organized 3-D oscillation. (d) Break-up: random, violent oscillations that end with the streak broken up into small motions.

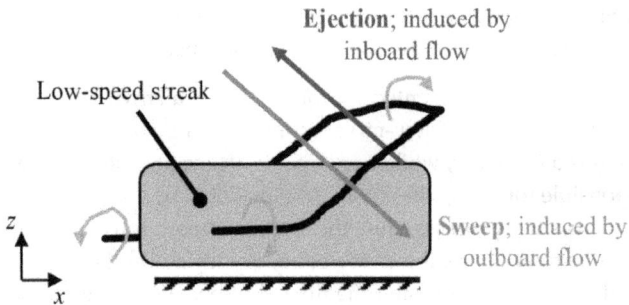

Figure 9. Ejection and sweep events caused by a hairpin vortex.

in the wall-pressure spectrum analysis held by Black [43]. Kim et al. [29] found that most/all occurrences of turbulent stresses $(-\overline{u'w'})$ take place not only in the near-wall region but entirely during the bursting process, which opts the bursting to be the main turbulence producer. This harmonizes with the early predictions of Runstadler et al. [44]. Kline et al. [37] anticipated the death of turbulence (*flow relaminarization*) when suppressing the bursting process and fetched many examples in this context:

- Relaminarizing turbulent boundary layer flow by applying a favorable pressure gradient; the pressure gradient hinders the lift-up process (a conclusion of the same paper [37] and other later, more comprehensive researches [45, 46]).

- Relaminarizing turbulent flow in a tube by rotating the tube about its axis; the centrifugal force affixes the streaks to the tube wall [47].

- Relaminarizing turbulent flow in a 2-D channel by rotating the channel about an axis fixed at one of the narrow walls and perpendicular to the mean flow direction; the Coriolis force suppresses turbulence at one wall and strengthens it at the other wall [48].

6. Streak generation

The streaks are created by streamwise vortices occupying the wall region [28, 49, 50]. Each streamwise vortex pumps the fast fluid from upper layers in one y direction and the slow fluid from wall-vicinity in the second direction. This action packs a body of high streamwise velocity at one side of the vortex and another of low streamwise velocity at the other side. These are termed the high-speed and low-speed streaks, see **Figure 10**.

For dye injected (or hydrogen bubbles generated) near the wall, the pumping action of the vortices accumulates the dye together with the wall-adjacent fluid in the low-speed streak. This is why the low-speed streaks appear in flow visualization experiments. It has been recorded that a streak can exist by its own [51], that is, the streamwise vortices form the streak and leave it behind. The streamwise vortices can be the legs of hairpin vortices. This means the streaks generate the hairpin vortices which in turn generate new streaks. This closes the turbulence self-sustenance cycle. The streak-hairpin-streamwise vortex mechanism is only one presumable mechanism for turbulence generation/maintenance among few others.

Robinson [10] and Schwartz [52] believed the low-speed streak to be lifted up or kinked by flow-induction from a streamwise vortex. Offen and Kline [53, 54] and Smith [42] have a somehow longer explanation. The vortical remnant from an upstream burst forms a traveling

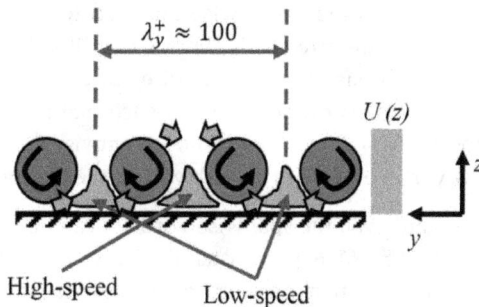

Figure 10. High- and low-speed streak generation by streamwise counter-rotating vortices.

pressure disturbance (instability). This traveling disturbance if passed over a low-speed streak impresses a local adverse pressure gradient upon a portion of it. This decelerates a part of the streak and hence lifts it up.

7. Generation of streamwise vortices

The streamwise, quasi-streamwise, vortices or rolls are the main turbulence producers in the viscous sublayer and responsible for low-speed streak formation. They are generated by different mechanisms:

1. They can be simply the legs of hairpin vortices. The legs of a hairpin vortex are quasi-streamwise and are usually attached to the wall i.e. lie in the sublayer. They range in length between 100 and 200 wall units [14]. There are two theories to interpret how these relatively short legs can produce the long streaks ($\Delta_x^+ \sim 1,000$). First, they sweep downstream along the wall, pack the streak and leave it behind in the long trails [55]. Second, many legs coalesce together and create the streak [21, 42, 56].

2. The streamwise vortices can be regenerated by other streamwise vortices [4, 14, 57, 58]. The shear (velocity gradient, $\frac{\partial U}{\partial z}$) causes the quasi-streamwise vortices to be stretched and lifted, that is, the upstream side is attached and the downstream side is detached from the wall (~9 ° inclination angle). Besides, the flow induced by other neighboring vortices tilts them in the spanwise direction (±4 °). The wall-normal detachment motion and spanwise tilt motion provoke high vorticity in the wall-normal direction. This wall-normal vorticity is then affected by the shear that stretches it and turns it in the streamwise direction. A child vortex is then born on the downwash side (flow toward the wall) at either the upstream or the downstream ends of the parent vortex. The direction of rotation of the child vortex is opposite to that of the parent vortex. The flow induced by the parent tilts the child in spanwise direction. The legs of a hairpin vortex can produce two pairs of streamwise vortices, inboard and outboard the hairpin [59]. Finally, we get a corrugated line of quasi-streamwise vortices.

3. The streamwise vortices can be generated during the bursting process [60]. Although the interaction between the low-speed streak and the mean flow produces a spanwise shear (vorticity) layer, this can turn in the streamwise direction to form a streamwise vortex. Depending on the presence of the streamwise vortex that lifts the streak, different types of vortices can be generated. If one lifting streamwise vortex is present at one side of the streak, the new vortex extends over it downstream such that the direction of rotation of the new vortex is opposing the old one, while if no streamwise vortices are present beside the streak, then the new vortex evolves in an arch vortex, see **Figure 11**.

4. The streamwise vortices, and even the spanwise vortices, can be produced by some low-speed streak *instability* [10, 50, 55, 61, 62]. An instability (waviness) in the spanwise direction can be excited by the asymmetric flanking-vortices. The waviness generates a streamwise vorticity layer. Once the waviness grows enough, it produces a strong velocity gradient in the streamwise direction, $\frac{\partial u}{\partial x}$. This gradient is responsible for stretching the aforementioned layer and collapsing (compressing) it into a streamwise vortex (circular vorticity

layer). Schoppa and Hussain [61] proposed three possible processes for vortex generation by streak instability—namely Process A: regeneration within gaps between consequent vortices; Process B: regeneration from an existing spanwise (arch) vortex, whose spanwise profile, excites streak instability to produce a pair of new streamwise vortices; and Process C: regeneration at trailing ends of low-speed streaks. Since the spawned vortices travel faster than the streak, they totally advect the streak leaving it behind and a new set of vortices are spawned.

5. Finally, streamwise vortices may generate from existing streamwise vorticity layers [63]. These vorticity layers were observed to evidence near the edge of the viscous sublayer. One layer tends to roll up into a compact streamwise core either due to the mutual induction with its image vorticity layer [50, 64] or by ejection from a parent, opposite signed, vortex [65]. The two mechanisms are illustrated in **Figures 12** and **13**, respectively.

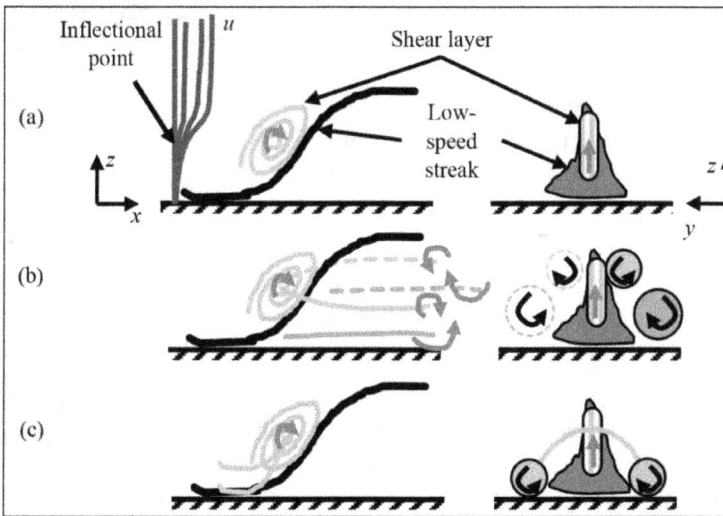

Figure 11. Generation of a streamwise vortex during the bursting process. (a) Lifted low-speed streak, (b) presence of the lifting streamwise vortex at one side of the streak, and (c) absence of the lifting vortex and generation of an arch vortex.

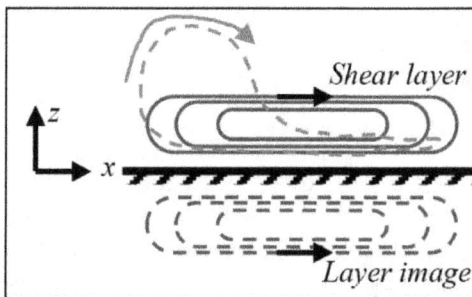

Figure 12. Roll-up of a streamwise shear layer by mutual induction with its image.

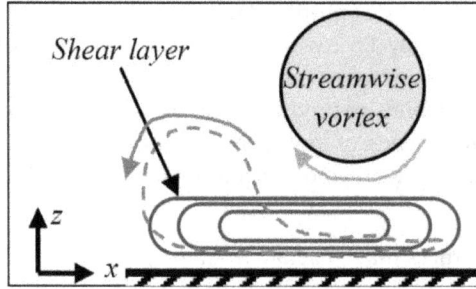

Figure 13. Roll-up of a streamwise shear layer by ejection from a parent vortex.

Either theory of the streamwise vortices implies that they are the main occupants of the viscous sublayer, whereas the outer layer is dominated by transverse vortices (heads of the hairpins). Bearing in mind their extended lengths compared to the spanwise (arch) vortices, the streamwise vortices are the main contributors to Reynolds stress in the sublayer [9, 27, 60, 66–71].

8. Turbulence sustenance by instability

In the foregoing discussion, the bursting process, and hence vortex generation, was almost totally undertaken by other coherent structures. This is termed the *parent-offspring* mechanism for turbulence production. On the contrary, a broad team of researchers designates a role in the vortex regeneration process to flow instabilities. The instabilities are supposed to take many forms and play different roles in turbulence generation [50, 72–74].

According to Swearingen and Blackwelder [75], instabilities motivate the generation of streamwise vortices and then trigger the generation of the hairpin vortices. Taylor-Görtler instabilities prevail near the wall due to streamline curvature, either as an inherent property of the TBL profile [73] or a result of the passage of a large-scale disturbance [36]. These can produce a system of streamwise vortices. The streamwise vortices in turn pump the fluid to build the low- and high-speed streaks. Consequently, two inflectional velocity profiles are formed ($u_{(z)}$ and $u_{(y)}$). From Rayleigh's criterion which has been upgraded by Fjørtoft's theorem, these inflectional profiles are inherently unstable [40, 55]. Thus, a secondary instability is generated which causes the streak lift-up, giving birth to new horseshoe vortices. Recall that the streak oscillations were interpreted by Kline et al. [37] as Kelvin-Helmholtz instabilities due to the growth and roll-up of the shear layer formed above the streak.

9. Large-scale motions (LSMs, vortex packets)

A vortex packet or large-scale motion is a bundle of hairpin vortices comprising 2–10 vortices aligned streamwise and traveling together. The inboard inductions of the hairpins against the mainstream combine together to form a relatively large region of uniform low momentum. The length of the packet ranges within 2–3δ. The recognition of vortex packets dates back to the early hairpin vortex visualizations [76]. According to Smith [42], the arrangement of the hairpins in packets is a natural consequence of their production as groups (2–5 vortices) in the bursting process. Zhou et al. [14] conducted a DNS to study the mechanism of generation of vortex packets. The simulation started with a pair of counter-rotating streamwise vortices which evolved into a hairpin. If this primary hairpin is strong enough, its induced flow interacts with the mean flow or induced flow from another hairpin to deliver secondary, tertiary, and downstream vortices. This vortex spawning complies with the findings of Doligalski et al. [77]. The final tent-like shape of the packet is very similar to the early hypothesis of Head and Bandyopadhyay [76].

The primary vortex and its offspring flock together as a packet. Adrian and his team [21, 78] further extended their *vortex packet paradigm* through experimentally studying the vortex packet behavior in the outer layer. The inboard induction of the packet hairpins against the main stream causes the packet to travel at a speed slower than the mean flow ($\sim0.8\,U_\infty$). As the hairpin ages, it expands in size and hence the induction is attenuated. Thus, the upstream parent hairpin moves faster than the downstream offspring; the packet stretches in the streamwise direction. Progressively, the overall induction of the packet hairpins is weakened. In addition, the hairpins move to higher faster fluid layers. Consequently, older packets move faster than younger packets and may overrun them. Packets can also merge with adjacent packets either streamwise or spanwise to form larger, stronger ones [79]. A vortex packet can extend to the edge of the boundary layer to form at least part of the turbulent bulges [3, 80]. Surprisingly, the description of the turbulent bulges introduced by Kovasznay et al. [81] agrees with that of the vortex packets. Moreover, Brown and Thomas [73] conducted correlation analysis across 75% of the TBL. They recognized large structures of length $2\,\delta$ and 18° inclination angle. From their PIV analysis, Ganapathisubramani et al. [82] confirmed the existence of the hairpin vortex packets. They could identify packets as long as $2\,\delta$. The packets hold more than 25% of Reynolds stress although occupying less than 4% of the total area. However, Ganapathisubramani et al. expected the packets to break down outside the logarithmic layer. While the angle of inclination of the hairpin vortex fluctuates around 45°, the packet as a whole leans against the wall at an angle of 10.5–13° [21, 78]. The LSMs are accompanied on either side with somehow shorter high-speed structures [13].

The vortex packet paradigm [21] assumes some kind of interaction between the large and small vortex packets. The larger packets move at higher speeds than the smaller ones such that they overtake them. As such, the smaller packets are liable to be enclosed in the uniform momentum zones of larger packets. As a consequence, the velocity vectors of the small scales undergo modulation by the larger ones. A modulating role for the large scales on the near-wall streaks was proven by Toh and Itano [83]. The modulation comprises the three velocity components [84] and extends to the frequency [85]. However, Hutchins [86] seizes the modulation to the near-wall region and interprets the similarity in amplitude between the scales outside it as a mere matter of preferential arrangement.

10. Very large scale motions (VLSMs, superstructures)

From their power spectral analysis of the streamwise velocity signal in channel and pipe flows, Jiménez [87] and Kim and Adrian [80] discovered a bimodal trend for the premultiplied spectrum. The two spectrum peaks correspond to wavelengths of 2–3 δ and 12–20 δ. The former was attributed to vortex packets (or turbulent bulges) whereas the latter was attributed to a turbulence coherent structure that extends very long streamwise. It was therefore named the very large scale motion (VLSM) or superstructure. Kim and Adrian conjectured that hairpins align in groups to form long LSMs, and LSMs in turn align coherently to form superstructures, see **Figure 14**. This hierarchical structure has been proven by Baltzer et al. [88] and received approbation from other authors [79, 89]. Nevertheless, Bailey et al. [90] found a disparity between the transverse (azimuthal) scales of both LSMs and superstructures that suggests the latter to be formed either by alignment of the biggest LSMs or separately from flow instabilities. Moreover, Hwang and Cossu [91–93] found from large eddy simulations (LESs) that the superstructures self-sustain even when the small-scale structures in the buffer and logarithmic layers are artificially quenched.

Superstructures were recorded over the lower half of the turbulent boundary layer, including the logarithmic region [89]. The superstructures contribute 50% to the turbulent kinetic energy and more than 50% to the Reynolds shear stress [94, 95]. Dennis and Nickels [89] conducted experimental (3D PIV + Talyor's hypothesis) tests on boundary layer flow from which they estimated the length of superstructures to be limited to $7\,\delta$. However, superstructures as large as $30\,\delta$ were found in pipe flow by DNS held by Lee and Sung [96] and by hotwire measurements held by Monty et al. [97]. The large difference between the two turbulent flows is caused by the free surface in case of TBL where entrainment occurs of large plumes from the free stream into the TBL. This entrainment breaks down the long coherent structures. Hutchins and Marusic [98] provided direct evidence of the superstructures in the logarithmic and lower wake regions of the turbulent boundary layer and atmospheric surface layer through velocity contours obtained from a rack of hotwires and sonic anemometers. Moreover, they found the superstructures to meander extremely along their length. The low-speed superstructures are usually twinned with high-speed structures of comparable lengths, probably induced by hairpin vortex legs [89]. The superstructure resembles an outer-layer counterpart of the low-speed streak [33]. The first is 12–$20\,\delta$ long and 3–$4\,\delta$ spanwise spaced, whereas the second is 1000 wall-units long and 100 wall-units spanwise spaced. Carlotti [99] differentiates between the two structures based on spectral analysis; the superstructures produce a "-1" power slope and the low-speed streaks produce a "-2" power slope.

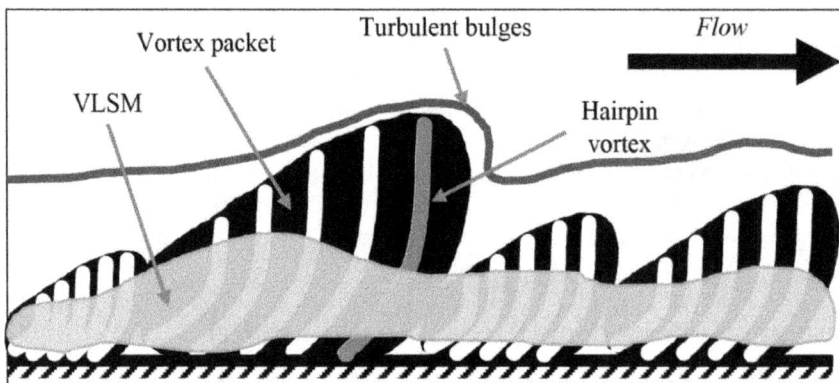

Figure 14. A VLSM turns out from agglomeration of several vortex packets aligned in the streamwise direction.

The newly discovered coherent structures (LSM and VLSMs) have been used to develop the old attached eddy model of Townsend [7]. Perry and co-authors [15, 18] devised a model to predict turbulence statistics by applying the attached eddy hypotheses to *a forest of hairpin vortices* of sizes proportional to their height from the wall. The model has been further polished by Marusic [100] who found the vortex packet to best resemble the attached eddies and prescribe turbulence statistics. Exactly the same was deduced by Dennis and Nickels [13]. Del Álamo et al. [101] inferred that the logarithmic region is populated with two classes of clusters; small detached vortex packets and tall attached packets. Hwang and Cossu [91–93] displayed that the energy-containing motions at a given spanwise length scale can self-sustain themselves by extracting energy directly from the mean flow even with the absence of any larger or smaller structures. They found the sizes of these energy-containing motions to be proportional to their distances from the wall, which makes them good candidates to be Townsend's attached eddies. However, they anticipated each of these eddies to be composed of two elements, a long streaky structure and a vortical structure. In the sublayer, these are the low-speed streak and the quasi-streamwise vortices flanking it and in the logarithmic and wake layers, these are the superstructures and the vortex packets aligning along them.

11. Generation of mechanical coherent structures in the ABL

The outer layer flow has generally been neutralized in the discussion about coherent structure generation. It was assumed that the structures are pure products of surface-instability interactions. This *bottom-up* model is convincing at low Re_τ where the inner layer resembles a considerable portion of the whole boundary layer depth. Nevertheless, as Re_τ increases, the inner and outer scales separate. For instance, the atmospheric boundary layer (ABL) can extend in height up to 1000 m but its inner layer is no more than few centimeters. It follows that, the bottom-up mechanism presupposes that a vortex packet of 5-cm size is to enlarge persistently tens of meters within a high Re_τ turbulent flow field until reaching the turbulent/non-turbulent interface.

Many authors [16, 17, 38, 40, 54, 73, 102, 103] recorded that the structures within the outer layer can trigger the bursting process, *top-down* models. Based on the synchronization between the bursting process and the passage of turbulent bulges in the turbulent/non-turbulent interface, Blackwelder and coworkers [28, 104, 105] conjectured that either the bursting phenomenon controls the outer flow field by developing the large-scale bulges, or else the outer field drives the bursts. Falco [106] observed the bulges at the turbulent/non-turbulent interface. He found the bulges to encompass *typical eddies*. These are ring or hairpin eddies found in almost all turbulent flows; wakes, jets, grid-generated turbulence; turbulent boundary layers, etc. In their proposed Overall Production Module, Falco, Klewicki, and Pan [107] hypothesized the bursting process to be triggered by a typical eddy moving toward the wall. The typical eddy when passing over a pair of low-speed streaks provokes the generation of two spanwise vortices within the streak pair, a primary vortex and a pocket vortex. The pocket in between opens up by self-induction and a sweep stream is created. Secondary hairpin vortices form across each streak. They are then twisted and rotated back toward the wall into the center of the pocket. This model largely coincides with the observations of Haidari and Smith [108].

Jiménez and Pinelli [62] utilized the capabilities of DNS to isolate the different turbulence sustenance mechanisms. They found that, at the studied Re_τ, turbulence can feed solely on the inner-layer cycle without any perturbations from the outer layer. However, Jiménez and Pinelli believe that the outer-layer perturbations still can maintain turbulence, yet at a lower activity. Between the bottom-up and top-down supporters, a third group of researchers [55, 109, 110] reckons that both models do coexist all the time with the former prevailing at low Re_τ and the latter at high Re_τ. Hunt and Morrison [103] suggest an Re_τ of 10^4 as a limiting value between the dominance regimes of both models.

Lin et al. [111] conducted an LES for neutral ABL. They employed a conditional sampling technique to track the evolution of the coherent structures. They concluded that hairpin vortices can be spawned by interaction between the ejection stream and either the mean flow or a sweep stream, a fact that has been confirmed later via DNS [14] and PIV [21]. The most interesting result among theirs is that they confirmed the top-down mechanism to generate vortices; a sweep stream when it impinges onto the ground, generates an ejection stream. However, they found these ejection-induced streams not to correlate with the strong ejection streams dominating the surface layer. Hunt and Morrison [103] and coworkers [99, 112, 113] developed the top-down model originally proposed by Falco [106] to comply with the ABL. Their conjecture is that the large eddies impinge and scrape along the surface, forming an internal boundary layer. As such, streamwise vortices of lengths several times the boundary layer height are generated alongside the impinging eddy. When the generated vortices interact with others, they are lifted far upward. The theory further splits the atmospheric surface layer into two sublayers: the shear layer and the eddy surface layer. The shear dominates the spectra in the first by distorting turbulence isotropy, while in the second the statistics are dominated by the ground-blocking effect on the impinging eddies (normal velocity suppression). They supported their theory by observations from atmospheric flow and spectrum analysis from the near surface region. The layer division was proven by spectral analysis of field measurements undertaken by [114]. Likewise, McNaughton and Brunet [115] postulated that the outer-layer eddies overtake the superstructures and induce hairpin vortices in a similar fashion to the near-wall cycle proposed by Kline et al. [37].

12. Conclusion

The turbulent flow stays as one of the most difficult scientific problems man has encountered. Despite the great deal of advance in the field, the path from the mean flow to the random fluctuations is still controversial. The modern experimental and numerical techniques are either one-eyed or biased toward the flow conditions synthesized by the researchers. This chapter reviewed the accumulated knowledge of TCSs and unfolded and compared their different mechanisms of generation. The scope was confined to turbulent boundary layer flow and atmospheric flow.

In boundary layer flows, turbulence is sustained by two concurrent mechanisms, the bottom-up mechanism and the top-down mechanism. The former dominates in low-Reynolds number (FPBL) flows and the latter dominates in high-Reynolds number (atmospheric) flows. The bottom-up mechanism generates turbulence coherent structures by surface-instability interaction, whereas the top-down mechanism relies on large outer-layer structures to trigger the generation process. Both the FPBL flow and the atmospheric flow share common features and are occupied by similar turbulence coherent structures, namely, the streamwise vortices, the low-speed streaks, the hair-pin vortices, the vortex packets, and the superstructures. However, the large scale in atmospheric flow neutralizes the role of the low-speed streaks and streamwise vortices. Many conceptual and numerical models have been set forth to enhance our understanding of turbulent flows. The research is always aiming to achieve a model that can be implemented in numerical simulations or drag reduction applications. In the end, despite the vast knowledge of turbulent flow structure, turbulence continues to be an unsolved or not thoroughly understood phenomenon.

Author details

Zambri Harun[1]* and Eslam Reda Lotfy[1,2]

*Address all correspondence to: zambri@ukm.edu.my

1 Faculty of Engineering and Built Environment, Universiti Kebangsaan Malaysia, Selangor, Malaysia

2 Mechanical Engineering Department, Alexandria University, Alexandria, Egypt

References

[1] Theodorsen T. Mechanism of turbulence. In: Proceedings of the Second Midwestern Conference on Fluid Mechanics. Vol. 1719; 1952

[2] Head MR, Bandyopadhyay P. New aspects of turbulent boundary-layer structure. Journal of Fluid Mechanics. 1981;**107**:297-338

[3] Adrian RJ. Hairpin vortex organization in wall turbulence. Physics of Fluids. 2007;**19**:41301

[4] Brooke JW, Hanratty TJ. Origin of turbulence-producing eddies in a channel flow. Physics of Fluids A: Fluid Dynamics. 1993;**5**:1011-1022

[5] Townsend AA. The Structure of Turbulent Shear Flow. Cambridge, New York; Cambridge University Press; 1956

[6] Grant HL. The large eddies of turbulent motion. Journal of Fluid Mechanics. 1958;**4**: 149-190

[7] Townsend AA. The Structure of Turbulent Shear Flow. Cambridge, New York: Cambridge Univ Press; 1976. p. 438p

[8] Willmarth WW, Tu BJ. Structure of turbulence in the boundary layer near the wall. Physics of Fluids. 1967;**10**:S134-S137

[9] Guezennec YG, Piomelli U, Kim J. On the shape and dynamics of wall structures in turbulent channel flow. Physics of Fluids A: Fluid Dynamics. 1989;**1**:764-766

[10] Robinson SK. Coherent motions in the turbulent boundary layer. Annual Review of Fluid Mechanics. 1991;**23**:601-639

[11] Klewicki J. Connecting vortex regeneration with near-wall stress transport. In: 29th AIAA, Fluid Dynamics Conference; 1998. p. 2963

[12] Smith CR, Walker JDA. Turbulent wall-layer vortices. Fluid Mechanics and its Applications. 1995;**30**:235

[13] Dennis DJC, Nickels TB. Experimental measurement of large-scale three-dimensional structures in a turbulent boundary layer. Part 1. Vortex packets. Journal of Fluid Mechanics. 2011;**673**:180-217

[14] Zhou J, Adrian RJ, Balachandar S, Kendall TM. Mechanisms for generating coherent packets of hairpin vortices in channel flow. Journal of Fluid Mechanics. 1999;**387**:353-396

[15] Perry AE, Henbest S, Chong MS. A theoretical and experimental study of wall turbulence. Journal of Fluid Mechanics. 1986;**165**:163-199

[16] Praturi AK, Brodkey RS. A stereoscopic visual study of coherent structures in turbulent shear flow. Journal of Fluid Mechanics. 1978;**89**:251-272

[17] Smith CR. Visualization of turbulent boundary layer structure using a moving hydrogen bubble wire probe. In: Lehigh Workshop on Coherent Structure in Turbulent Boundary Layers; 1978. pp. 48-49

[18] Perry AE, Chong MS. On the mechanism of wall turbulence. Journal of Fluid Mechanics. 1982;**119**:173-217

[19] Lozano-Durán A, Jiménez J. Time-resolved evolution of coherent structures in turbulent channels: Characterization of eddies and cascades. Journal of Fluid Mechanics. 2014;**759**:432-471

[20] Brinkerhoff JR, Yaras MI. Numerical investigation of the generation and growth of coherent flow structures in a triggered turbulent spot. Journal of Fluid Mechanics. 2014;**759**:257-294

[21] Adrian RJ, Meinhart CD, Tomkins CD. Vortex organization in the outer region of the turbulent boundary layer. Journal of Fluid Mechanics. 2000;**422**:1-54

[22] Wallace JM, Eckelmann H, Brodkey RS. The wall region in turbulent shear flow. Journal of Fluid Mechanics. 1972;**54**:39-48

[23] Wilcox DC et al. Turbulence Modeling for CFD. Vol. 2. La Canada, CA: DCW Industries; 1998

[24] Corrsin S. Some current problems in turbulent shear flows. In: Symposium on Naval Hydrodynamics. Vol. 515; 1957

[25] Ferrell JK, Richardson FM, Beatty KO Jr. Dye displacement technique for velocity distribution measurements. Industrial and Engineering Chemistry. 1955;**47**:29-33

[26] Alfonsi G. Coherent structures of turbulence: Methods of eduction and results. Applied Mechanics Reviews. 2006;**59**:307-323

[27] Blackwelder RF, Eckelmann H. Streamwise vortices associated with the bursting phenomenon. Journal of Fluid Mechanics. 1979;**94**:577-594

[28] Blackwelder RF. The bursting process in turbulent boundary layers. In: Lehigh Workshop on Coherent Structure in Turbulent Boundary Layers; 1978, pp. 211-227

[29] Kim H, Kline SJ, Reynolds WC. The production of turbulence near a smooth wall in a turbulent boundary layer. Journal of Fluid Mechanics. 1971;**50**:133-160

[30] Schraub FA, Kline SJ. A Study of the Structure of the Turbulent Boundary Layer with and without Longitudinal Pressure Gradients. Stanford University; 1965

[31] Tu BJ, Willmarth WW. An Experimental Study of Turbulence near the Wall through Correlation Measurements in a Thick Turbulent Boundary Layer. Ann Arbor: Dept. Aerosp. Eng., University of Michigan; 1966. Technical Report No. 02920-3-T

[32] Smith CR, Metzler SP. The characteristics of low-speed streaks in the near-wall region of a turbulent boundary layer. Journal of Fluid Mechanics. 1983;**129**:27-54

[33] Pujals G, Garc'ia-Villalba M, Cossu C, Depardon S. A note on optimal transient growth in turbulent channel flows. Physics of Fluids. 2009;**21**:15109

[34] Del Álamo JC, Jimenez J. Linear energy amplification in turbulent channels. Journal of Fluid Mechanics. 2006;**559**:205-213

[35] Butler KM, Farrell BF. Optimal perturbations and streak spacing in wall-bounded turbulent shear flow. Physics of Fluids A: Fluid Dynamics. 1993;**5**:774-777

[36] Hamilton JM, Kim J, Waleffe F. Regeneration mechanisms of near-wall turbulence structures. Journal of Fluid Mechanics. 1995;**287**:317-348

[37] Kline SJ, Reynolds WC, Schraub FA, Runstadler PW. The structure of turbulent boundary layers. Journal of Fluid Mechanics. 1967;**30**:741-773

[38] Willmarth WW. Structure of turbulence in boundary layers. Advances in Applied Mechanics. 1975;**15**:159-254

[39] Corino ER, Brodkey RS. A visual investigation of the wall region in turbulent flow. Journal of Fluid Mechanics. 1969;**37**:1-30

[40] Grass AJ. Structural features of turbulent flow over smooth and rough boundaries. Journal of Fluid Mechanics. 1971;**50**:233-255

[41] Nakagawa H, Nezu I. Structure of space-time correlations of bursting phenomena in an open-channel flow. Journal of Fluid Mechanics. 1981;**104**:1-43

[42] Smith CR. A Synthesized Model of the Near-Wall Behavior in Turbulent Boundary Layers. Bethlehem Pennsylvania: Department of Mechanical Engineering and Mechanics, Lehigh University; 1984

[43] Black TJ. An analytical study of the measured wall pressure field under supersonic turbulent boundary layers. National Aeronautics and Space Administration. Washington D.C.; 1968

[44] Runstadler PW, Kline SJ, Reynolds WC. An experimental investigation of the flow structure of the turbulent boundary layer. Stanford University, California: 1963

[45] Harun Z, Monty JP, Mathis R, Marusic I. Pressure gradient effects on the large-scale structure of turbulent boundary layers. Journal of Fluid Mechanics. 2013;**715**:477-498

[46] Blackwelder RF, Kovasznay LSG. Large-scale motion of a turbulent boundary layer during relaminarization. Journal of Fluid Mechanics. 1972;**53**:61-83

[47] Cannon JN. Heat transfer from a fluid flowing inside a rotating cylinder [thesis]. Stanford University; 1965

[48] Halleen RM, Johnston JP. The Influence of Rotation on Flow in a Long Rectangular Channel: An Experimental Study. California: Stanford University, Thermosciences Division; 1967

[49] Eitel-Amor G, Örlü R, Schlatter P, Flores O. Hairpin vortices in turbulent boundary layers. Physics of Fluids. 2015;**27**:25108

[50] Schoppa W, Hussain F. Coherent structure generation in near-wall turbulence. Journal of Fluid Mechanics. 2002;**453**:57-108

[51] Jeong J, Hussain F. On the identification of a vortex. Journal of Fluid Mechanics. 1995;**285**: 69-94

[52] Schwartz SP. Investigation of vortical motions in the inner region of a turbulent boundary layer [thesis]; 1981

[53] Offen GR, Kline SJ. A proposed model of the bursting process in turbulent boundary layers. Journal of Fluid Mechanics. 1975;**70**:209-228

[54] Offen GR, Kline SJ. Combined dye-streak and hydrogen-bubble visual observations of a turbulent boundary layer. Journal of Fluid Mechanics. 1974;**62**:223-239

[55] Panton RL. Overview of the self-sustaining mechanisms of wall turbulence. Progress in Aerospace Science. 2001;**37**:341-383

[56] Dennis DJC. Coherent structures in wall-bounded turbulence. Anais da Academia Brasileira de Ciências. 2015;**87**:1161-1193

[57] Bernard PS, Thomas JM, Handler RA. Vortex dynamics and the production of Reynolds stress. Journal of Fluid Mechanics. 1993;**253**:385-419

[58] Miyake Y, Ushiro R, Morikawa T. The regeneration of quasi-streamwise vortices in the near-wall region. JSME International Journal Series B: Fluids and Thermal Engineering. 1997;**40**:257-264

[59] Zhang N, Lu L, Duan Z, Yuan X. Numerical simulation of quasi-streamwise hairpin-like vortex generation in turbulent boundary layer. Applied Mathematics and Mechanics. 2008;**29**:15-22

[60] Heist DK, Hanratty TJ, Na Y. Observations of the formation of streamwise vortices by rotation of arch vortices. Physics of Fluids. 2000;**12**:2965-2975

[61] Schoppa W, Hussain F. Genesis of longitudinal vortices in near-wall turbulence. Meccanica. 1998;**33**:489-501

[62] Jiménez J, Pinelli A. The autonomous cycle of near-wall turbulence. Journal of Fluid Mechanics. 1999;**389**:335-359

[63] Sendstad O. The near wall mechanics of three-dimensional turbulent boundary layers. Stanford University, California; 1992

[64] Jiménez J, Orlandi P. The rollup of a vortex layer near a wall. Journal of Fluid Mechanics. 1993;**248**:297-313

[65] Orlandi P. Vortex dipole rebound from a wall. Physics of Fluids A: Fluid Dynamics. 1990;**2**:1429-1436

[66] Kim J, Moin P, Moser R. Turbulence statistics in fully developed channel flow at low Reynolds number. Journal of Fluid Mechanics. 1987;**177**:133-166

[67] Bakewell HP Jr, Lumley JL. Viscous sublayer and adjacent wall region in turbulent pipe flow. Physics of Fluids. 1967;**10**:1880-1889

[68] Robinson SK. The kinematics of turbulent boundary layer structure. NASA STI/Recon Technical Report N. 1991;**91**:26465

[69] Clark JA, Markland E. Vortex structures in turbulent boundary layers. Aeronautical Journal. 1970;**74**:243-244

[70] Clark JA, Markland E. Flow visualization in free shear layers. Journal of the Hydraulics Division. 1973;**99**:1897-1913

[71] Cantwell BJ. Organized motion in turbulent flow. Annual Review of Fluid Mechanics. 1981;**13**:457-515

[72] Benney DJ. A non-linear theory for oscillations in a parallel flow. Journal of Fluid Mechanics. 1961;**10**:209-236

[73] Brown GL, Thomas ASW. Large structure in a turbulent boundary layer. Physics of Fluids. 1977;**20**:S243-S252

[74] Phillips WRC, Wu Z, Lumley JL. On the formation of longitudinal vortices in a turbulent boundary layer over wavy terrain. Journal of Fluid Mechanics. 1996;**326**:321-341

[75] Swearingen JD, Blackwelder RF. The growth and breakdown of streamwise vortices in the presence of a wall. Journal of Fluid Mechanics. 1987;**182**:255-290

[76] Head MR, Bandyopadhyay P. Combined flow visualization and hot wire measurements in turbulent boundary layers. In: Smith CR, Abbott, editors. Lehigh Workshop on Coherent Structure in Turbulent Boundary Layers; 1978. pp. 98-129

[77] Doligalski TL, Smith CR, Walker JDA. Vortex interactions with walls. Annual Review of Fluid Mechanics. 1994;**26**:573-616

[78] Christensen KT, Adrian RJ. Statistical evidence of hairpin vortex packets in wall turbulence. Journal of Fluid Mechanics. 2001;**431**:433-443

[79] Lee JH, Sung HJ. Very-large-scale motions in a turbulent boundary layer. Journal of Fluid Mechanics. 2011;**673**:80-120

[80] Kim KC, Adrian RJ. Very large-scale motion in the outer layer. Physics of Fluids. 1999;**11**:417-422

[81] Kovasznay LSG, Kibens V, Blackwelder RF. Large-scale motion in the intermittent region of a turbulent boundary layer. Journal of Fluid Mechanics. 1970;**41**:283-325

[82] Ganapathisubramani B, Longmire EK, Marusic I. Characteristics of vortex packets in turbulent boundary layers. Journal of Fluid Mechanics. 2003;**478**:35-46

[83] Toh S, Itano T. Interaction between a large-scale structure and near-wall structures in channel flow. Journal of Fluid Mechanics. 2005;**524**:249-262

[84] Talluru KM, Baidya R, Hutchins N, Marusic I. Amplitude modulation of all three velocity components in turbulent boundary layers. Journal of Fluid Mechanics. 2014;**746**:R1

[85] Baars WJ, Talluru KM, Hutchins N, Marusic I. Wavelet analysis of wall turbulence to study large-scale modulation of small scales. Experiments in Fluids. 2015;**56**:188

[86] Hutchins N. Large-scale structures in high Reynolds number wall-bounded turbulence. Progress in Turbulence V. Berlin: Springer; 2014, p. 75-83

[87] Jiménez J. The largest scales of turbulent wall flows. CTR Annual Research Briefs. 1998;**137**:54

[88] Baltzer JR, Adrian RJ, Wu X. Structural organization of large and very large scales in turbulent pipe flow simulation. Journal of Fluid Mechanics. 2013;**720**:236-279

[89] Dennis DJC, Nickels TB. Experimental measurement of large-scale three-dimensional structures in a turbulent boundary layer. Part 2. Long structures. Journal of Fluid Mechanics. 2011;**673**:218-244

[90] Bailey SCC, Hultmark M, Smits AJ, Schultz MP. Azimuthal structure of turbulence in high Reynolds number pipe flow. Journal of Fluid Mechanics. 2008;**615**:121-138

[91] Hwang Y, Cossu C. Self-sustained process at large scales in turbulent channel flow. Physical Review Letters. 2010;**105**:44505

[92] Cossu C, Hwang Y. Self-sustaining processes at all scales in wall-bounded turbulent shear flows. Philosophical Transactions of the Royal Society A. 2017;**375**:20160088

[93] Hwang Y. Statistical structure of self-sustaining attached eddies in turbulent channel flow. Journal of Fluid Mechanics. 2015;**767**:254-289

[94] Guala M, Hommema SE, Adrian RJ. Large-scale and very-large-scale motions in turbulent pipe flow. Journal of Fluid Mechanics. 2006;**554**:521-542

[95] Jimenez J, Del Alamo JC, Flores O. The large-scale dynamics of near-wall turbulence. Journal of Fluid Mechanics. 2004;**505**:179-199

[96] Lee JH, Sung HJ. Comparison of very-large-scale motions of turbulent pipe and boundary layer simulations. Physics of Fluids. 2013;**25**:45103

[97] Monty JP, Stewart JA, Williams RC, Chong MS. Large-scale features in turbulent pipe and channel flows. Journal of Fluid Mechanics. 2007;**589**:147-156

[98] Hutchins N, Marusic I. Evidence of very long meandering features in the logarithmic region of turbulent boundary layers. Journal of Fluid Mechanics. 2007;**579**:1-28

[99] Carlotti P. Two-point properties of atmospheric turbulence very close to the ground: Comparison of a high resolution LES with theoretical models. Boundary-Layer Meteorology. 2002;**104**:381-410

[100] Marusic I. On the role of large-scale structures in wall turbulence. Physics of Fluids. 2001;**13**:735-743

[101] Del Álamo JC, Jimenez J, Zandonade P, Moser RD. Self-similar vortex clusters in the turbulent logarithmic region. Journal of Fluid Mechanics. 2006;**561**:329-358

[102] Rao KN, Narasimha R, Narayanan MAB. The "bursting"phenomenon in a turbulent boundary layer. Journal of Fluid Mechanics. 1971;**48**:339-352

[103] Hunt JCR, Morrison JF. Eddy structure in turbulent boundary layers. European Journal of Mechanics - B/Fluids. 2000;**19**:673-694

[104] Blackwelder RF, Kaplan RE. On the wall structure of the turbulent boundary layer. Journal of Fluid Mechanics. 1976;**76**:89-112

[105] Blackwelder R. An experimental model for near-wall structure. In: 29th AIAA, Fluid Dynamics Conference; 1997. p. 2960

[106] Falco RE. Coherent motions in the outer region of turbulent boundary layers. Physics of Fluids. 1977;**20**:S124-S132

[107] Falco RE, Klewicki JC, Pan K. Production of turbulence in boundary layers and potential for modification of the near wall region. Structure of Turbulence and Drag Reduction. Berlin: Springer; 1990, p. 59-68

[108] Haidari AH, Smith CR. The generation and regeneration of single hairpin vortices. Journal of Fluid Mechanics. 1994;**277**:135-162

[109] Hutchins N, Chauhan K, Marusic I, Monty J, Klewicki J. Towards reconciling the large-scale structure of turbulent boundary layers in the atmosphere and laboratory. Boundary-Layer Meteorology. 2012;**145**:273-306

[110] Mathis R, Hutchins N, Marusic I. Large-scale amplitude modulation of the small-scale structures in turbulent boundary layers. Journal of Fluid Mechanics. 2009;**628**:311-337

[111] Lin C-L, McWilliams JC, Moeng C-H, Sullivan PP. Coherent structures and dynamics in a neutrally stratified planetary boundary layer flow. Physics of Fluids. 1996;**8**:2626-2639

[112] Högström U, Hunt JCR, Smedman A-S. Theory and measurements for turbulence spectra and variances in the atmospheric neutral surface layer. Boundary-Layer Meteorology. 2002;**103**:101-124

[113] Hunt JCR, Carlotti P. Statistical structure at the wall of the high Reynolds number turbulent boundary layer. Flow, Turbulence and Combustion. 2001;**66**:453-475

[114] Drobinski P, Carlotti P, Newsom RK, Banta RM, Foster RC, Redelsperger J-L. The structure of the near-neutral atmospheric surface layer. Journal of the Atmospheric Sciences. 2004;**61**:699-714

[115] McNaughton KG, Brunet Y. Townsend's hypothesis, coherent structures and Monin–Obukhov similarity. Boundary-Layer Meteorology. 2002;**102**:161-175

The Turbulence Regime of the Atmospheric Surface Layer in the Presence of Shallow Cold Drainage Flows: Application of Laser Scintillometry

John A. Mayfield and Gilberto J. Fochesatto

Additional information is available at the end of the chapter

http://dx.doi.org/10.5772/intechopen.80290

Abstract

The presence of shallow cold flows in the atmospheric boundary layer (ABL) instigates changes in the turbulent regime of the atmospheric surface layer (ASL). This small scale flow circulation introduces radiative cooling controls over large areas in polar latitudes during winter. In this study, microscale dynamic and turbulent variables have been obtained in the framework of the Winter Boundary Layer Experiment in Fairbanks, Alaska, developed during the winters of 2009/2010 and 2010/2011. Multiscale surface turbulence observations based on Eddy covariance and laser scintillometry were combined with Doppler acoustic sounding to document simultaneous changes in the ABL flow and ASL turbulence. We computed changes in momentum and heat fluxes characterizing intermittent and persistent modes of the drainage flow over three study cases. On the basis of laser scintillometry observations, we argue that a significant source of turbulence aiming at the surface fluxes has origins in the upper level shear-induced thermal turbulence at the top of the ABL.

Keywords: surface turbulence, optical scintillation, radiative cooling, small scale drainage flows, stable boundary layer

1. Introduction

The absence of daylight combined with snow-covered surfaces during the extreme winters in continental Alaska sets up a unique meteorological condition in the lower troposphere where surface radiative cooling becomes the dominant forcing mechanism initiating the formation of stably-stratified ABL [1, 2]. This meteorological feature is present under synoptic anticyclone conditions (i.e., surface high pressure, clear skies) and is particularly prominent and episodic throughout the winter. In general, when this meteorological configuration sets up, several days of a persistent stably-stratified ABL occur [3–6]. This local ABL feature, also known as surface-based temperature inversion (SBI), occurs statistically over 85% of time in the presence of multiple elevated temperature inversion (EI) layers from synoptic origin [1]. In such anti-cyclone conditions and when weak horizontal synoptic flow forcing prevails, the topographic configuration and the orientation of mountains in the Interior of Alaska constrain the low-level tropospheric circulation. At the regional scale, the air flow in the ABL becomes locally quasi-laminar and regionally stagnant under strong surface radiative cooling and therefore prone to disruptions from local-scale circulation mechanisms. The investigation of this process was the objective of the Winter Boundary-Layer Experiment [7].

Mountains and hills in polar regions shelter large cold pools of air between them that are exposed to a higher radiative cooling rate than adjacent low-laying areas that may be more opened and oriented toward south (i.e., the south-sunny side). These topographic and radiative conditions set the stage for the development of density flows that takes the form of drainage flows [7]. Such small-scale flows channeled through the varying terrain morphology connect basins on the north-facing to those on the south-facing sides of the mountains and hills and profoundly affect the stability of the relatively lower and warmer basins, as well as the surface turbulent flux and momentum regime. The occurrence of these flows has been identified in other experiments on lower latitudes [8, 9] as a cold down-slope winds initiated by the temperature contrast between basins influencing therefore the strongly stratified ABL that builds up locally at the basin scale.

On the other hand, extreme winter conditions in Fairbanks, Alaska, have been known for several decades to represent a very complex air pollution problem [3, 10, 11]. Besides local anthropogenic emissions, what exacerbates the air pollution problem is precisely the naturally occurring strong surface cooling rate and, consequently, the formation of extremely low-level SBI, often complicated with multiple stratified layers [1, 2] under a stagnant flow condition. As indicated previously, these episodes are mostly present under specific synoptic conditions (e.g., calm winds and stagnant quiescent synoptic anticyclonic flows) [3, 5] over periods lasting several days [1]. Moreover, the meteorological setup, the flow configuration, and the complexity of the developed ABL structure impose stringent modeling restrictions when an accurate representation of the ABL is needed in particular for air pollution assessments [12].

The first observational study of the winter ABL in Fairbanks was carried out in mid-1970s using a bi-static sodar instrument from the National Oceanic and Atmospheric Administration, Wave Propagation Laboratory. The set of observations illustrated several dynamic aspects of the stable ABL such as propagation of waves, inertial oscillations within the ABL structure, and formation and destruction of temperature inversion and stratified layers [5, 6]. Despite of the importance of these observations in terms of improving the understanding of physical processes in the ABL and the outstanding mesoscale modeling challenges that winter high latitudes imposes no additional efforts have been dedicated to continuing these investigations until the development of Wi-BLEx [1, 2, 7]. In fact, previous studies [10, 13–15] have indicated that during the extreme winter, a heavy air mass lies close to the ground, where SBI forms as the result of a strong surface radiative cooling rate pronounced by the absence of sunlight and under a stagnant airflow often with wind speed less than ~1 ms⁻¹. In such conditions, katabatic flows have been found above the local SBI as they come down from the mountains surrounding the area. In these previous studies, and perhaps with limited observational capabilities at the time, it was also indicated that the extreme cold pool dominating the Tanana Valley air mass prevented the local SBI airflow to mix with the upper level katabatic flow. This feature is self-evident in several photographs of Fairbanks winter scenery, where a clear differentiation can be visually established when comparing tall power plant emission stacks and home heating emission systems. In this study, we provide observational evidence that during the winter period, the valley ABL is penetrated by a small-scale cold flow from the north-northwest sector of the observational site (see **Figure 1**) close to the hill slopes around Fairbanks. This shallow cold flow was observed to interact with the local ABL under specific prescribed stagnant anticyclone conditions and to dominate the thermodynamic structure and circulation of the ABL close to the foothills for specific periods of time [7]. In this case, penetrative drainage flows were observed to introduce small scale mixing, increasing thermal turbulence near the top of the ABL and re-stratification at the surface layer when the flow ceased. This flow originates upstream in sheltered cold pools of air (i.e., northern slopes of the Cranberry Hills), where the combination of geographic orientation and very low solar elevation angle during the winter allows for a more efficient radiative cooling than in the central valley region, i.e., the south facing slopes of the Tanana Valley (see **Figure 1**).

Section 2 describes the experimental setup during Wi-BLEx, and Section 3 describes the large scale meteorological conditions during the observations. Section 4 introduces the observations of drainage flows and ABL structure, and Section 5 analyzes and discusses the cases and

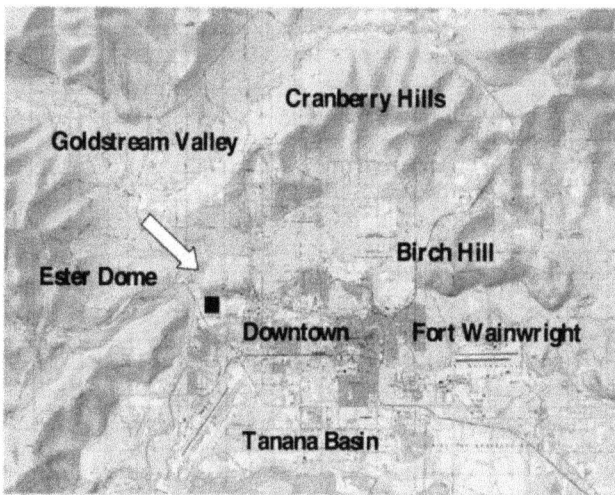

Figure 1. Topographic map of the Fairbanks area interior of Alaska. Black square indicates experimental site on the campus farm of the University of Alaska Fairbanks. The arrow indicates the direction of flow into the experimental site, where Wi-BLEx was developed. (Source: Alaska seamless USGS topographic maps).

the determination of structural and turbulent parameters of the drainage flow that influence the state of surface turbulent fluxes. This chapter concludes in Section 6 where a summary conclusion is provided based on the experimental evidence that supports the hypothesis that shear-induced thermal turbulence in the upper level depth of the drainage flow propagates downward to induce large-eddy turbulent exchanges in the ASL fluxes.

2. Experiment and methodology

Wi-BLEx took place at the University of Alaska Fairbanks (UAF) experimental farm located in the south-facing basin in the foothills of the Cranberry Hills (see **Figure 1**) on the UAF campus. The specific sets of observations discussed in this paper were carried out from November 1, 2010, to March 30, 2011, and three cases were selected for further analysis.

The instrumentation consisted of two 3D-sonic anemometers installed at 2- and 4-m height on a meteorological mast around the center of the farm (see **Figure 2**).

The sonic anemometers acquired at 10 Hz, the three components of the velocity field, as well as the sonic virtual temperature. This allows calculating turbulent quantities u', v', w', and θ'. Surface-layer observations were then combined with data from an ABL profiler. In this case, an acoustic phased-array Doppler sodar (Remtech PA-0) was installed 130 m north of the sonic anemometer tower (**Figure 2**, left panel). Sodar profiles ranged from 20 to 500 m on average and were able to determine the coefficient of turbulent temperature structure parameter C_T^2, wind speed, wind direction, and vertical velocity at 10-min time intervals and 10-m vertical resolution. In order to complement the set of observations, a large aperture scintillometer (LAS) (Scintec BLS 900) was set

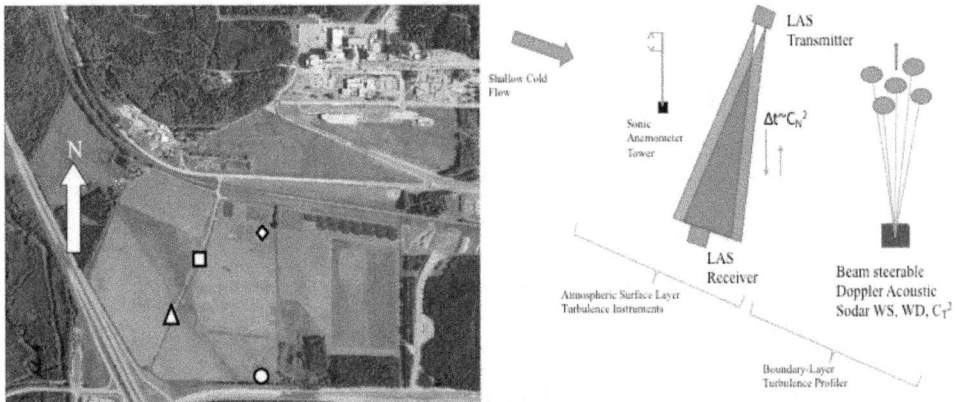

Figure 2. Aerial photo of the UAF campus farm (left panel) and conceptual idea illustrating instruments and specific observations (right panel). The site extends 1.2 km west-to-east and 700 m north-to-south. The instruments were located ~400 m from major aerodynamic perturbations upstream to the northwest caused by low-level forested area. The triangle represents the micrometeorological tower, the rhombus represents the LAS receiver, the circle is the LAS emitter, and the square is the location for the Doppler acoustic sodar. Right panel illustrate the variables and instruments applied to retrieve ASL and ABL turbulence variables (source: Photo courtesy of Alan Tonne UAF farm manager).

across the basin north-south orientation to acquire the structure coefficient of refractive index turbulence C_N^2. The instrument's layout was superimposed upon an aerial photo of the UAF Campus Farm in **Figure 2** left panel, while **Figure 2** right panel shows the instrument-measurement concept. The sonic anemometers were factory calibrated before being placed in the field. Previous to the field deployment, the anemometers were installed in a closed chamber, and the acquisition electronic was calibrated for the zero-speed velocity measurement on the three components of the wind speed and temperature against the Väisälä temperature and relative humidity probe HMT-330. Similarly, the sampling rate of the sonic anemometer and the high-speed data logger were controlled for sampling rate synchronization and correct time-stamp acquisition. The methodology used to signal-process turbulent data sets from sonic anemometers and LAS follows nonlinear median filters for despiking signals [16]. And, earlier to calculating turbulent quantities, high-pass filtering techniques were applied based on the streamline coordinate rotation method [17, 18]. However, unlike the general case of turbulent flux determination in micrometeorology, the time interval to apply this methodology and preserve the turbulent fluctuations of the drainage flow events was determined to be over a window of 5 min. This time-window allows achieving the best compromise between sampling error of turbulent magnitudes and the representation of the wind measurements in terms of minimizing the probability of change in the wind direction [19]. The temperature data were processed using the same time-window interval (5 min) and using a low-pass autoregressive moving average digital filter to extract the low frequency component [20].

Based on the retrieved turbulent magnitudes, the combined auto-covariance of wind speed components and their cross-covariance allowed calculation of turbulent kinetic energy (tke) Eq. (1), surface friction velocity (u*) Eq. (2), surface momentum (τ_0) Eq. (3), and the flux Richardson number (Ri_f) Eq. (4).

$$tke = \frac{1}{2}[\overline{(u')^2} + \overline{(v')^2} + \overline{(w')^2}]$$
(1)

$$u^* = \left(\overline{(u'w')}^2 + \overline{(v'w')}^2\right)^{\frac{1}{4}}$$
(2)

$$\tau_0 = \overline{u'v'}$$
(3)

$$Ri_f = \frac{-\frac{g}{\theta_{sfc}}\overline{w'\theta'}}{\overline{u'w'}\frac{dU}{dz} + \overline{v'w'}\frac{dU}{dz}}$$
(4)

$$C_N^2 = 1.22 \, D^{\frac{7}{3}} d^{-3} \sigma_{ln(I)^2}$$
(5)

In Eqs. (1)–(4), θ_{sfc} is the surface potential temperature and U(z) is the wind profile and in Eq. (5), D is the diameter of the receiver's optical lens, d is the distance between emitter and receiver, while $\sigma_{ln(I)}^2$ is the variance of the natural logarithm of the recorded optical intensity over a given time period. The LAS emitter was operated at 125 Hz and 1 min averaging in the receiver at 520 m distance across the basin (see **Figure 2**). The acquired data were processed to calculate C_N^2 according to Eq. (5) at 1 min time period [21]. Temperature and pressure data were acquired at approximately the center of the LAS beam and were used to estimate the heat fluxes based on Monin-Obukhov similarity hypothesis corrected by the stability functions depending of the Obukhov stability parameter (z/L) [22, 23] over a flat, snow-covered surface [24].

3. Meteorological and mean flow conditions during observations

Three cases were selected for analysis from the set of experiments carried out during Wi-BLEx. In this section, the synoptic meteorology framework and the mean flow conditions are described.

Case I (January 18–19, 2011): A surface high-pressure system was centered over the central Yukon and overtime built a surface pressure strength rising to 1040 hPa. A surface high (1043 hPa) over Siberia maintained the anticyclone motion of air masses in the interiors of Alaska. The ABL flow in Fairbanks was in stagnant condition under a weak pressure gradient force in the region, resulting in weak winds and clear skies which drove a large radiation cooling of the basins nearby the observational area. **Figure 3** illustrates the wind direction on the left panel and the wind speed on the right panel over the site. These plots are a combination of ASL and upper ABL level measured by sonic anemometer 4 m above the surface and sodar observations at 55 m and 165 m in the ABL and the free atmosphere (FA), respectively.

Case II (February 7, 2011): Observations were carried out under the influence of a surface high-pressure system that covered the Yukon and central Alaska regions (~1037 hPa surface

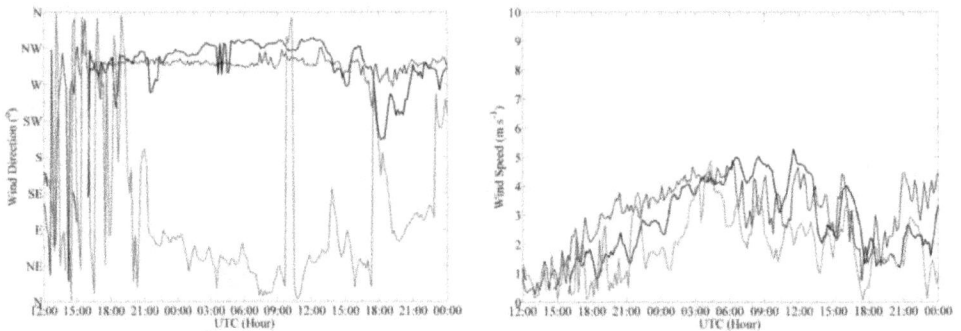

Figure 3. January 18–19 mean wind direction (left panel) and wind speed evolution (right panel) on the ASL, ABL layers, and FA. Left panel: Black solid line represents 10 min average wind direction measured by sonic anemometer 4 m height, the black dash line represents the ABL wind speed at 55 m measured by the sodar, and the gray line represents 165 m measured by the sodar. Right panel: Black solid line represents 10 min average wind speed in the ASL measured by sonic anemometer 4 m height, the black dash line represents the ABL wind speed at 55 m measured by the sodar, and the gray line represents the 165 m measured by the sodar.

pressure). The pressure gradient force was weak over the Interior of Alaska, resulting in low ABL winds at the observational site. Low-level cloud coverage was suppressed by the presence of the anticyclone in the region. The surface layer flow verified the occurrence of nighttime drainage events in sequences of 270 min, followed by a drainage event of 120 min and five subsequent short-period intermittent flows in decreasing wind speed each one lasting less than 60 min. In **Figure 4,** panel on the left illustrates the wind direction and panel on the right depicts the wind speed over the site. These plots are a combination of surface 4 m height sonic anemometer 10 min averaged wind direction and wind speed and Doppler sodar in upper levels in the ABL and the FA.

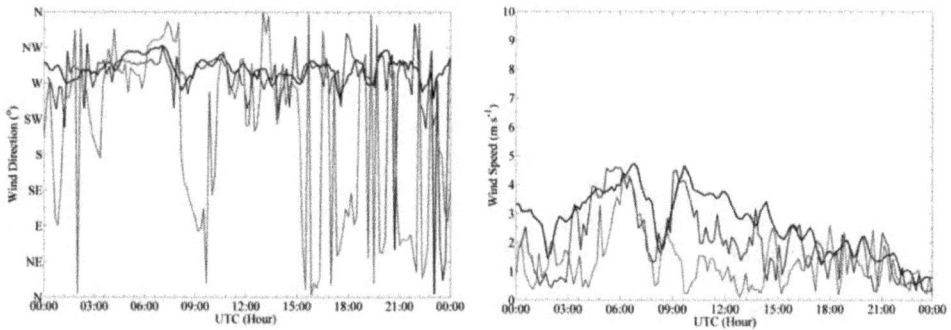

Figure 4. February 07, 2011. Left panel is the mean wind direction and right panel is the wind speed evolution in the ASL, ABL, and FA. Left panel: Black solid line represents 10 min average wind direction measured by sonic anemometer 4 m height, the black dash line represents the ABL wind speed at 55 m measured by the sodar, and the gray line represent the 165 m measured by the sodar. Right panel: Wind speed in the ASL and in the ABL layers. Black solid line represents 10 min average wind direction measured by sonic anemometer 4 m height, the black dash line represents the wind speed at 55 m measured by sodar in the ABL, and the gray line represent the 165 m measured by sodar in the FA.

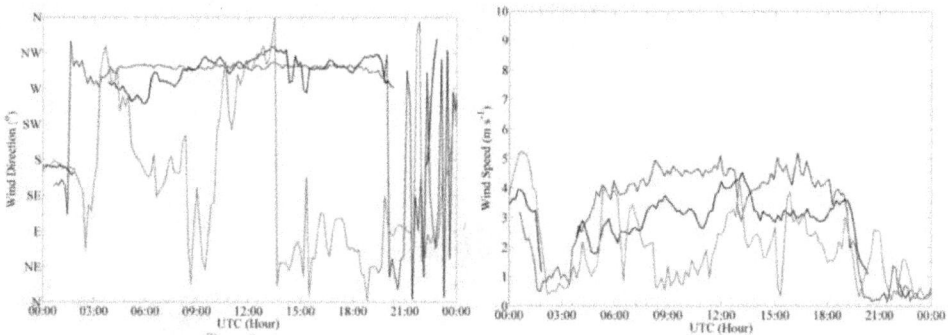

Figure 5. March 6, 2011. Left panel is the mean wind direction and right panel is the wind speed evolution in the ASL, ABL layers, and FA. In the left panel, black solid line represents 10 min average wind direction measured by sonic anemometer 4 m height, the black dash line represents the ABL wind speed at 55 m measured by the sodar, and the gray line represent the 165 m measured by the sodar. In the right panel, black solid line represents 10 min average wind speed measured by sonic anemometer 4 m height, the black dash line represents the ABL wind speed at 55 m measured by the sodar, and the gray line represents the 165 m measured by the sodar in the FA.

Case III (March 6, 2011): Observations during late winter verified the presence of a surface ridge to the south ~40 km from the observational site with tight isobars over southeast Alaska. Surface high pressure (~1036 hPa) centered over the Alaska/Yukon border established a pressure gradient force that weakened as the surface high pressure moved over the Brooks Range. The ABL state in the Tanana Valley was forced by the high pressure located around the southern Yukon and the ridge to the southwest. The winds at the geostrophic level were from the west and northwest during the night, while during the day, they were from the east-southeast. In **Figure 5**, panel on the left illustrates the wind direction and the panel on the right depicts the wind speed over the site. Similar to **Figures 3** and **4**.

4. Drainage flow events and changes in the surface turbulence

Case I: Surface wind direction observations showed a constant northwest wind direction between 15:00 UTC on 18 January and 15:00 UTC on 19 January as described in **Figure 3**. At the beginning of the event, the temperature in the atmospheric surface layer was ~−22°C and is observed to decrease in 3 h to ~−29°C as depicted in **Figure 6** left panel; vertical velocity (w) slowly increases, central panel, as the drainage flow starts to penetrate the ABL turning the surface wind direction from the northwest on the right panel and increasing surface winds.

Significant changes are observed in the surface turbulent regime obtained by auto-covariance of turbulent velocities and cross-covariances to calculate tke, u^*, and τ_0. These variables are shown in **Figure 7** where in the right panel is the tke ($m^2 s^{-2}$), central panel the increase of $u^*(ms^{-1})$ and right panel is the momentum (m^2s^2). At the beginning, previous to the initiation of the drainage flow, the surface layer static stability was maintained to 1°C/m of stratification. During drainage flow penetration (15:00 to 21:00 UTC), the tke increased to 0.15 $m^2 s^{-2}$, u^* increased to 0.15 ms^{-1}, and τ_0 increased to 0.02 m^2s^2. The surface wind speed increased to 3.5 ms^{-1}, and the temperature dropped down by ~4°C. At 00:00 UTC, the temperature steadily decreased from −27.5°C to −30°C. When the drainage flow ceased, the surface layer re-stabilize as indicated by the calculation of the Ri_t number.

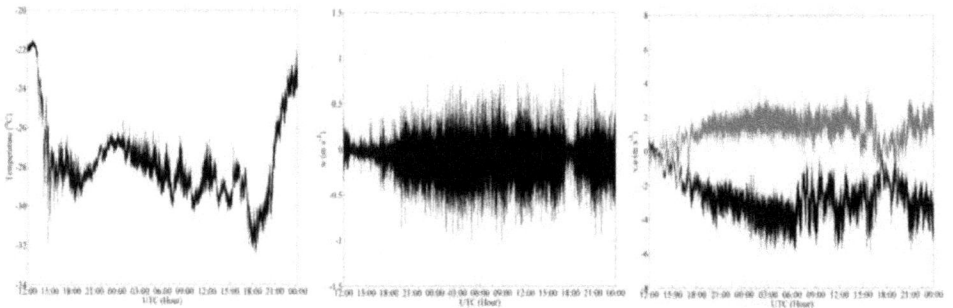

Figure 6. Turbulent variables for Case I period January 18–20, 2011, measured at 10 Hz, 4 m height sonic anemometer. Left panel represents the sonic temperature; central panel represents the vertical velocity; and right panel are the components u (black trace) and v (gray trace) of the horizontal wind speed.

The vertical structure of the ABL from the 00:00 UTC 18 January shows a surface-based inversion at 108 m from the radiosonde measurements at the NWS-NOAA nearby station with winds from the south in the stable ABL and the FA [25]. This height can also be retrieved in the backscatter signature of the sodar profile by searching the point of maximum gradient indicating the decaying of thermal turbulence structure in the ABL by the C_T^2 (see **Figure 8**). Vertical and time variabilities of the C_T^2 reveal changes introduced in the ABL structure by the presence of the drainage flow on 18 January at ~15:00 UTC. The flow lasted until 19 January at 15:00 UTC. The ABL wind speed in the period from 15:00 to 22:30 UTC did not exceed 2 ms^{-1}, but in the second period from 22:30 to 17:30 UTC on 19 January, the wind speed exceeded 3 ms^{-1}. During the first period, the drainage flow was 100 m deep into the ABL, while during second period, it was shallower ~95 m. **Figure 8** displays the ABL vertical structure measured

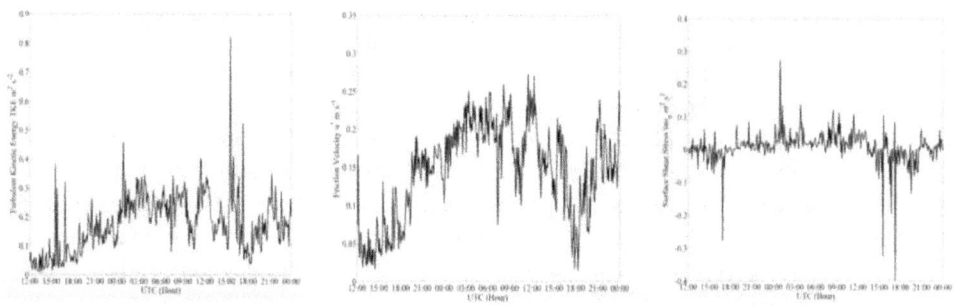

Figure 7. Case I. ASL turbulent parameters for Case I at 4 m height tke (left panel), friction velocity (central panel) and surface momentum (right panel).

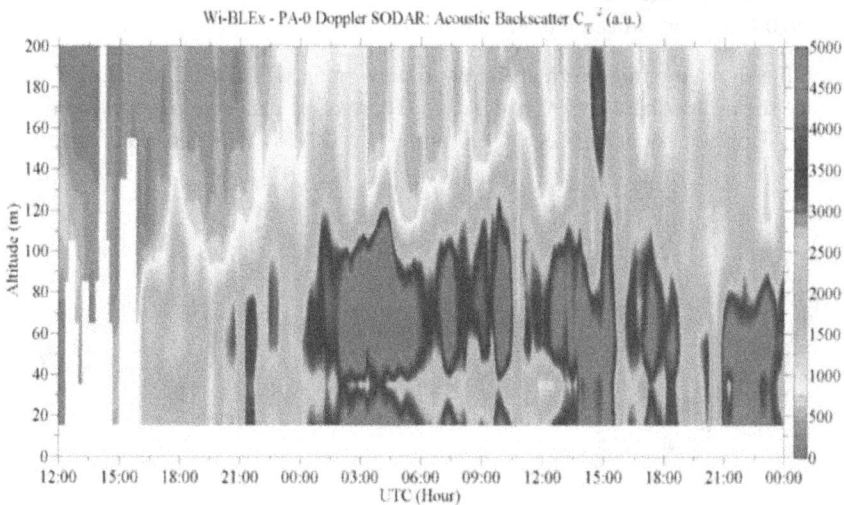

Figure 8. Case I. vertical structure of the ABL from 18 to 20 January, 2011. The sodar represents the thermal turbulent structure coefficient C_T^2 in arbitrary units with time resolution of 10 min and vertical resolution of 10 m.

by sodar, from 20 to 200 m, during the occurrence of the drainage flow. The color scale indicates an increase of the thermal turbulence C_T^2 overtime.

The C_T^2 coefficient in **Figure 8** displays a variable structure in space-time indicating increasing thermal turbulence regime as the drainage flow establishes and dominates the flow in the ABL. The ASL displays a perturbation in turbulence regime covering from surface up to ~40 m height. While this turbulence reappears after 40 m up to a range between 80 and 120 m, the spatial structures observed in the ABL responded differently to the drainage flow penetration because of the vertical stratification in the stable ABL and the increasing level of surface friction. Turbulent structures within the ABL and at the ASL overlap overtime in the period after 12:00 to 15:00 and between 21:00 and 00:00 and split for about 6 hours before 06:00 to 12:00. Remarkably, for some short periods of time, the profiler displays a complete disappearance of the thermal turbulence in the ABL for example from 19:00 to 20:00.

Case II: The beginning of this period occurred at sunset around 01:20 UTC 7 February where the initial ABL state indicated an already well-established surface-based inversion. The 00:00 UTC radiosonde from the NWS-NOAA radiosonde station PAFA showed a temperature inversion layer up to 98 m [25]. The Doppler sodar showed ABL winds from the west and FA winds from the southeast in **Figure 4**. The vertical structure measured by Doppler sodar identified a drainage flow penetrating the ABL of the basin at 4:40 UTC. The flow irruption to the basin lasted until 07:30 UTC. Wind direction observations at the surface showed an intermittent flow from northwest wind direction between 00:00 UTC on 7 February and 00:00 UTC on 8 February as described in **Figure 4**. At the beginning of the event, the temperature in the surface layer was ~−8°C and is observed to decrease in 3 h to ~−24°C as depicted in **Figure 9** (left panel), while vertical velocity (w) slowly increases **Figure 9** (central panel) as the drainage start to penetrate the ABL turning the surface wind direction from the northwest **Figure 9** (right panel). The τ_0 increased from 0 to 0.04 m²s² as well as the tke from 0 to 0.1 m² s⁻² and u* from 0.05 to 0.2 ms⁻¹, as indicated in **Figure 10** (left, central, and right panels).

The C_T^2 profile showed a minimum at the wind maximum, while the C_T^2 maximum occurred at the wind shear layer between 90 and 110 m. The second drainage penetration event was observed by Doppler sodar between 09:10 and 11:10 UTC with a vertical structure of the

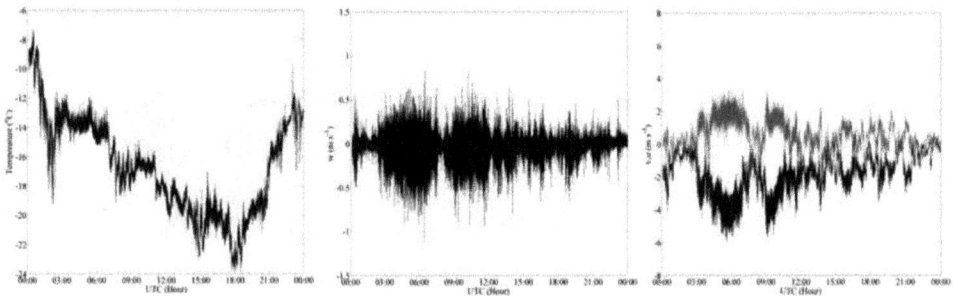

Figure 9. Case II. Turbulent variables for February 7, 2011, measured at 10 Hz, 4 m height sonic anemometer. Left panel represents the sonic temperature; central panel represents the vertical velocity; and right panel are the components u (black trace) and v (gray trace) of the horizontal wind speed.

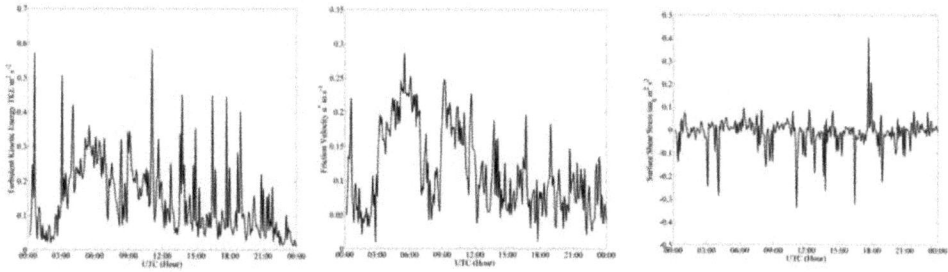

Figure 10. Case II. ASL turbulent parameters at 4 m height tke (left panel), friction velocity (central panel) and surface shear stress (right panel).

flow (i.e., wind profile) approximately similar to that of the first event. In total, this drainage episode was characterized by seven intermittent short-time duration of small scale flows. The first event lasted ~270 min. The wind speed in the surface layer was maintained at 4 ms^{-1}. This flow broke up around 07:30 UTC with a change in surface wind direction from the southwest and decreasing wind speed to less than 1 ms^{-1}. The second drainage flow event occurred from 09:00 to 12:00 UTC. The first half of this period sustained a surface wind speed that reached 4 ms^{-1}, but wind speed decreased to 2 ms^{-1} during the second half of this event. Similarly, τ_0 increased from approximately 0 to 0.03 m^{2}s^2, the tke increased to ~0.03 m^2 s^{-2}, and u* increased to 0.02 ms^{-1}. The five remaining events lasted less than 30 min developing surface wind speed less than 2 ms^{-1}. Calculation of surface turbulent magnitudes retrieved similar behavior as compared to Case I, despite this case is composed by short bursts of sub-mesoscale flows. During the five intermittent pulses in the latter half of the day, strong stratification aloft occurred, indicated by the vertical C_T^2 in **Figure 11**. **Figure 10** shows the changes in the surface turbulence parameters, tke increased to 0.1 m^2 s^{-2}, and momentum increased to 0.03 m^{2}s^2. The vertical structure of the ABL (see **Figure 11**) increases in C_T^2 as result of the increasing temperature variations. In the upper ABL levels, the drainage flow introduces localized levels of turbulence bounded by stratify layers from 75 m in 125 min with similar behavior in five other intermittent pulses that occurred with wind speed less than 2 ms^{-1} at 13:30, 15:20, 16:20, 19:10, and 20:30 UTC. The Doppler sodar shows evidence of the drainage pulse with an increase in horizontal wind speed and increased and localized C_T^2 values from 20 to 120 m.

Case III: In this case, the sunset on the March 6, 2011 occurred at 03:20 UTC and the surface layer was in near-neutral state; flows in the FA were from the south-southeast, and wind speeds at the 400 to 500 m level exceeded 8 ms^{-1}. The surface temperature was −9.5°C, and the surface stratification reached 3°C between the two sonic anemometer levels (see **Figure 12**). The ABL begin a rapid stratification after ceasing solar radiation input, giving rise to a rapid radiative cooling at the surface and air layers aloft. The drainage flow penetrated the ASL with the early nocturnal inversion in the time interval from 03:40 to 03:50 UTC. Surface winds turned to the northwest direction between 06:10 to 06:20 UTC. Later on, the surface winds displayed a northwest to west-northwest wind direction between 04:30 and 19:30 UTC. When the flow events ceased, the static stability buildup immediately as the wind speed slowed down to less than 1 ms^{-1} at the surface and tke, u*, and τ_0 dropped to very low values as shown in **Figure 13**.

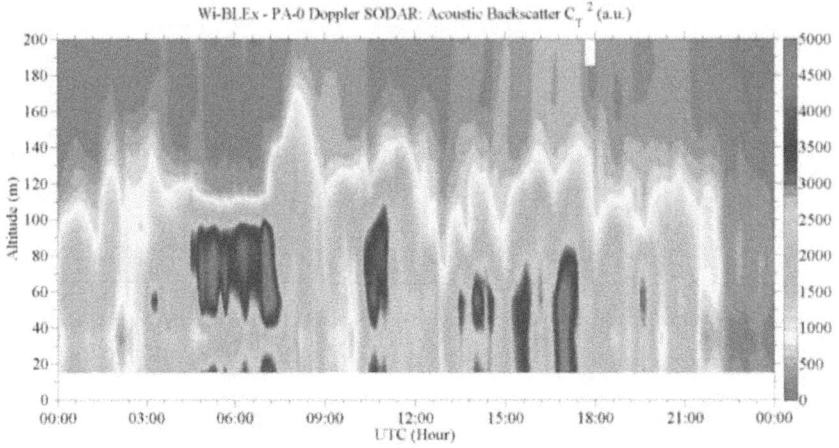

Figure 11. Case II. Vertical structure of the ABL during February 7, 2011, represented by the thermal turbulent structure coefficient C_T^2 in arbitrary units with time resolution of 10 min and vertical resolution of 10 m.

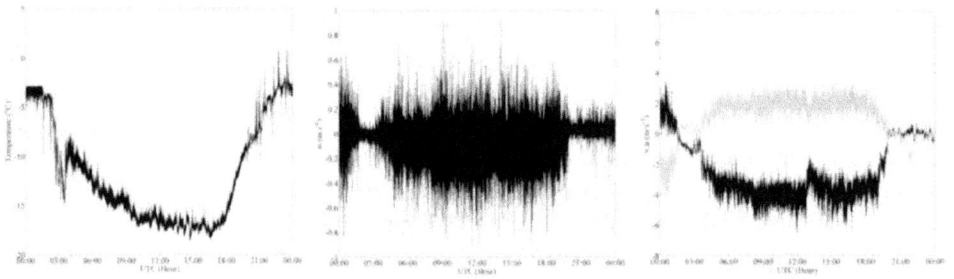

Figure 12. Case III. ASL turbulent flow measured at 10 Hz, 4-m height sonic anemometer during March 6, 2011. Left panel is the sonic temperature. Central panel is the vertical velocity and right panel are the components u (black trace) and v (gray trace) of the horizontal wind speed.

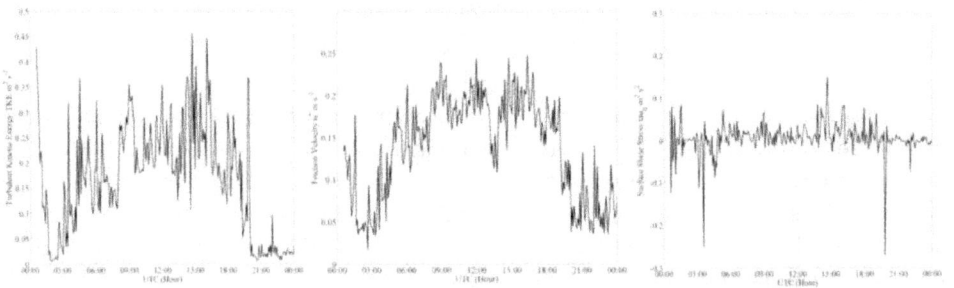

Figure 13. Case III. ASL turbulent parameters at 4 m. Left panel represents tke, central panel represents friction velocity, and right panel is the surface momentum.

During this period of time, the surface layer was in a transitional period until the vertical cooling initiated the formation of a stable ABL. As the initial drainage flow from the northwest direction penetrated the surface layer, the tke increased from ~0 to 0.2 m² s⁻², u, increased from ~0.05 to 0.2 ms⁻¹, and τ_0 increased from ~0 to 0.03 m² s⁻², while the temperature decreased from −3°C to −18°C, a drop in about 15°C in ~15 h.

The Ri_f calculated was highly variable taking negative and positive values before the onset of the drainage flow. Ri_f decreased and reached a steady value above 1 corresponding to a quasi-laminar flow. The end of the disruptive flow was indicated by the change of wind direction to the southeast and a reduction in the wind speed and the surface micrometeorological variables tke, u*, and τ_0 at 20:00 UTC. The sun rose at 16:21 UTC, and the surface-layer temperature began to rise at 17:00 UTC. The most active turbulent activity was verified to occur between 09:00 and 15:00 UTC, when the surface parameters were at their maximum dynamic swing and wind speed was sustained overtime, while the Ri_f number indicates a dynamic unstable flow. The vertical ABL structure observed by Doppler sodar indicates increasing C_T^2 values as shown in **Figure 14** from 09:00 to 15:00 UTC when, at the same time, the wind speed was accelerating in the vertical. The $C_T{}^2$ within the region decreased suddenly in the layer from 50 to 100 m, while at the surface, high fluctuations occurred in the Ri_f number and $\tau_{0'}$ and the wind speed decreased (see **Figure 5**). The flow re-stabilized afterward and became dynamically stable with Ri_f number ~2. After 18:00 UTC, the Ri_f number variability increased as the flow stability broke down. The vertical profile of C_T^2 fluctuated at higher values during the transition of change in wind direction followed by an increased stratification in the basin. Of note here, **Figure 12** (central panel) illustrates a significant drop in the turbulent vertical velocity.

The vertical C_T^2 structure of the stable ABL assumed a characteristic pronounced parabolic shape up to the top of the stable ABL when the drainage penetrates. The C_T^2 profiles obtained from the Doppler sodar show a particular structure, with higher values in the vertical range from 20 to 60 m and a minimum at the drainage flow wind speed maximum. A second maximum produced the strongest signal in the acoustic echo and represents a shear layer at 100-m height. The depth of the drainage flow varied between 80 and 120 m depth. This shear layer induced thermal turbulence and entrainment of air into the ABL. In the time interval between 09:00 and 15:00 UTC, the drainage flow wind speed accelerated from 3 to 5 ms⁻¹ with an increase of the $C_{T'}^2$ reaching a maximum peak wind speed of 6 ms⁻¹. Between 18:00 and 21:00 UTC, the flow decelerated from 5 ms⁻¹ with an increase C_T^2 (See **Figure 14**).

Summarizing, records of the ABL vertical structure obtained from Doppler sodar observations in **Figures 8** and **11–14** indicate the presence of localized turbulence in upper levels of the ABL when the drainage flow is present in the basin; in this case represented by an increases of C_T^2. Scrutinizing further the data sets obtained by Doppler sodar (i.e., wind speed and wind direction), it was verified that this turbulence appears to be induced by shear mechanism on layers along the side of the drainage flow. Similar to results from [26–28] who relates the sodar scattering cross-section to vertical potential temperature gradient and wind shear, [29] and [28] showed an echogram facsimile in which drainage flow velocity and the echo intensity increases. Nevertheless, in polar atmospheres during winter, this signature unequivocally represents shear instead of convective plume development. This differentiation was

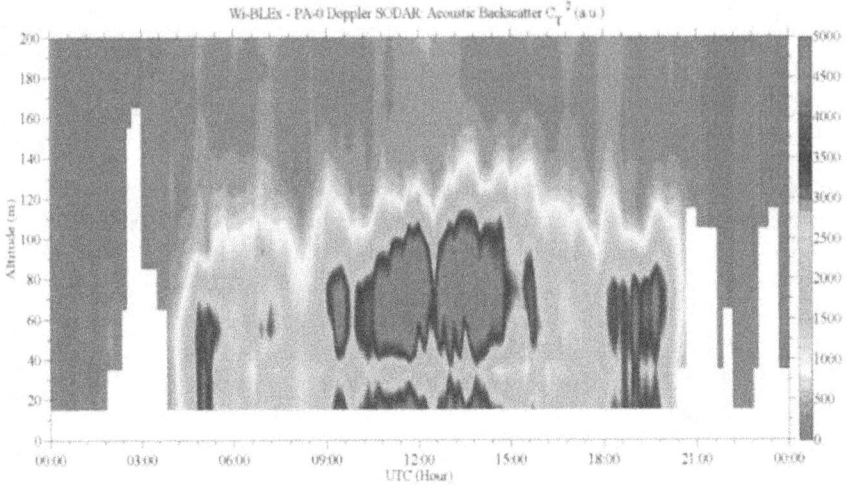

Figure 14. Case III. Vertical structure of the ABL from 6 March sodar represented by the thermal turbulent structure coefficient C_T^2 in arbitrary units with time resolution of 10 min and vertical resolution of 10 m.

illustrated for very different atmospheric and surface conditions by [30]. In this early study, the signature of acoustic backscattering in the presence of drainage flow and shear driven by convective plumes that arise from surface heating was clearly demonstrated.

5. Discussion

The turbulent state of the ASL was observed to significantly change in the presence of the shallow cold drainage flow into the basin. After scrutinizing all recorded cases during Wi-BLEx, it was found that this flow develops based on two modes: persistent flow represented by Case I and III and intermittent flow represented by Case II. Changes in the turbulence regime of the ASL, instigated by the penetration of the drainage flow into the basin's ABL, have been determined by means of two instruments sonic anemometer and LAS. These two instruments sample basically the same turbulence spectrum with the difference that sonic anemometers are in-situ sensors, while LASs are large-scale area-average turbulence sensing devices. Both instruments respond to the turbulence developed by the flow upstream on overlapping footprints. However, in this experiment, the LAS was installed across the basin (see **Figure 2**) to continuously evaluate the turbulent state of the drainage flow in space and time and thus fully record the turbulent structures developing from microscale to basin scale.

Based on Monin-Obukhov similarity hypothesis, the sensible heat flux was calculated for both instruments LAS (H_{LAS}) and sonic anemometer as indicated previously using 5-min intervals eddy-covariance integration (H_{EC}) [23, 24]. The H_{LAS} was calculated using friction velocity (u*) and the Obukhov length (L) obtained based on sonic anemometers measurements. The LAS covered an optical path length of 520 m across the basin. In this case, H_{EC} was proven to capture in the range of 70–80% up to 100% of the surface turbulent fluxes developing at the basin scale when compared to H_{LAS} [7]. This result, indicating a divergence between the calculated heat fluxes over certain time periods, pointed to the idea that large-eddy turbulence could be present in the ASL [31, 32]. On the other hand, it can be argued that given the variability of the flow in the ASL, a time variable integration is needed to account for the entire eddy flux for the case of H_{EC}. Nevertheless, the nature and source of the large eddies present in the basin can be independently investigated. These eddies can be either part of the natural mode by which the drainage flow develops breaking up the stratification in the basin (i.e., by carrying large eddy momentum) or they can be eventually part of the turbulence generated at the ABL level that would breakdown and dissipate at the surface. To further speculate on the nature of the large-eddy inducing optical turbulence in the LAS system, **Figure 15** conceptualizes the turbulent transfer mechanism supporting the argument that shear-induced turbulence along the side of the drainage flow enhances C_T^2 in the ABL and propagates down to the surface as seen by LAS.

Therefore, in what follows we analyze, each study case based on LAS optical scintillation times series and the calculated spectrogram. This analysis is guided by the flow dynamic and turbulence (i.e., wind direction, speed, and the C_T^2) register by Doppler sodar and sonic anemometers (section 3 and 4). In this case, the use of surface and ABL profiler allows determining the time interval when the drainage flow breaks into the basin and also evaluates the intervals when large eddies are present in the surface due to turbulence in ABL levels.

In the analysis of Case I, it was noted that between 15:00 and 00:00 UTC of the following day, at the initiation of the drainage flow, H_{EC} on average is ~–4 Wm^{-2}, while H_{LAS} is ~–8 Wm^{-2} without major flux differences. However, in the period from 00:00 to 09:00 UTC, the time averaged turbulent flux values are H_{LAS} –18 Wm^{-2} and H_{EC} –14 Wm^{-2}, showing an increasing divergence between the measurements. This discrepancy is consistent with the increasing probability of wind direction change thus affecting the H_{EC} measurements. Because of the dynamic changes in the flow, the H_{EC} and H_{LAS} measurements are therefore not convergent during this period (see **Figures 3** and **6**).

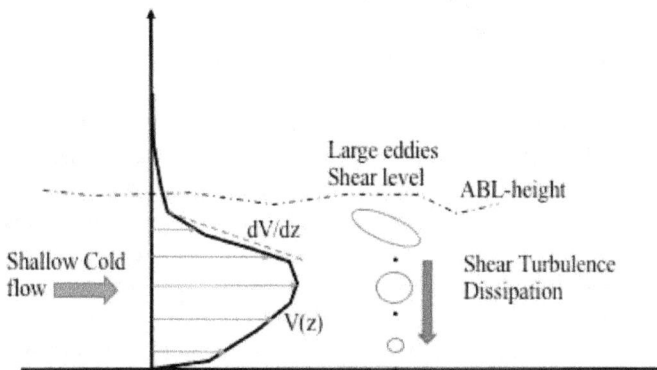

Figure 15. Conceptual scheme describing the mechanism for shear induced turbulence in the presence of a shallow cold flow. Upper ABL level illustrates wind speed shear that breaks down propagating and dissipating toward the surface. These localized large eddies impact the optical turbulence signature of LAS.

The time-series of C_N^2 exhibits large excursions in the order of ~2 to 4.10^{-13} $(m^{-2/3})$ mainly at the beginning and at the end of the study case from 12:00 to 18:00 January 18 and after 18:00 on January 19. This time variability can be seen in **Figure 16** top panel. However, during the time period where the flow in the basin establishes and fully develops, as depicted in **Figures 3** and **6**, the optical turbulence of refractive index verifies a steady increase in the signal level. This smooth increase in turbulence level corresponds to the signature described in **Figure 4**, where an increase in the turbulent intensity in w' is verified. Mostly as the flow establishes and increases speed, it also develops high-frequency turbulence as depicted in **Figure 6** which in turn also develops further tke and u*. However, the time series of C_N^2 (**Figure 16** top panel) also verifies sudden signal increases lasting longer time periods basically from ~ 10 min to more than 1 h between 18:00 and 15:00 UTC of the following day. Therefore, in order to analyze the observed variability in C_N^2 in terms of time-frequency contribution to the turbulence spectrum and confirm the existence of localized low-frequency turbulence, a multiresolution analysis based on continuous wavelet transform was conducted as depicted in **Figure 16** bottom panel. Scrutinizing this spectrogram in more detail, a sustained level of turbulence is shown to appear fluctuating overtime on periods going from 32 min. to less than 256 min. During this period of time, the flow is established in the basin given the levels of wind speed, preserved wind direction, raise of u* and tke, and therefore one of the sources for large eddies could be the localized turbulence appearing in the ABL's upper levels as illustrated by Doppler sodar in **Figure 8**. This dynamic connection is facilitated by the fact that the ASL stratification has been erased by the drainage flow penetration. This spectrogram also clearly depicts the existence of large-eddies in the basin at the beginning of the drainage flow period 12:00 to 18:00 and when the flow ceased to occur after 18:00. However, in these two time periods, the source of large eddy turbulence is basically the flow momentum transport and its horizontal variability. In fact, during these two periods of time, the flow in the basin meanders and in the first case by gaining momentum breaks the stratification propagating large eddies in the basin, while in the second period, the flow ceases meandering carrying less-energetic large eddies.

For Case II on 7 February 2011, LAS was not working for about 30 min. after 21:00 UTC. Thus, in order to provide a continuous spectral analysis, the signal was cut-off at 21:00 UTC as seen in **Figure 17** both panels. In this case, the average H_{LAS} and H_{EC} reached values of ~ −20 Wm^{-2} with the first sustained drainage flow pulse, while the second intermittent flow pulse developed an H_{LAS} and H_{EC} on average ~−16 Wm^{-2}. The wind speed in the two intermittent pulses changes rapidly over a short period of time, and eddy covariance was calculated over 5 min. integration period as indicated previously. LAS and EC flux measurements compared well in time. Similarly, to the analysis of Case I, the time series of C_N^2 exhibits periods of time with strong turbulent development reaching values of 4.10^{-13} m$^{-2/3}$ signal with a much higher time variability and sustained turbulence levels for periods of hours (**Figure 17**, top panel). This time variability in the surface turbulence is given by the flow intermittency that manifests on burst of C_N^2 lasting a couple of hours. The spectrogram in **Figure 16** (bottom panel) exhibits several periods in which spectral deposition of turbulence is verified at time scales from 16 to 256 min due to the presence of large eddies aiming the flow in the basin. This spectral patchiness correlates in some cases with localized turbulence occurring in upper shear layers of the ABL as demonstrated by localized enhancements of C_T^2 (see **Figure 11**). However, unlike Case I, the

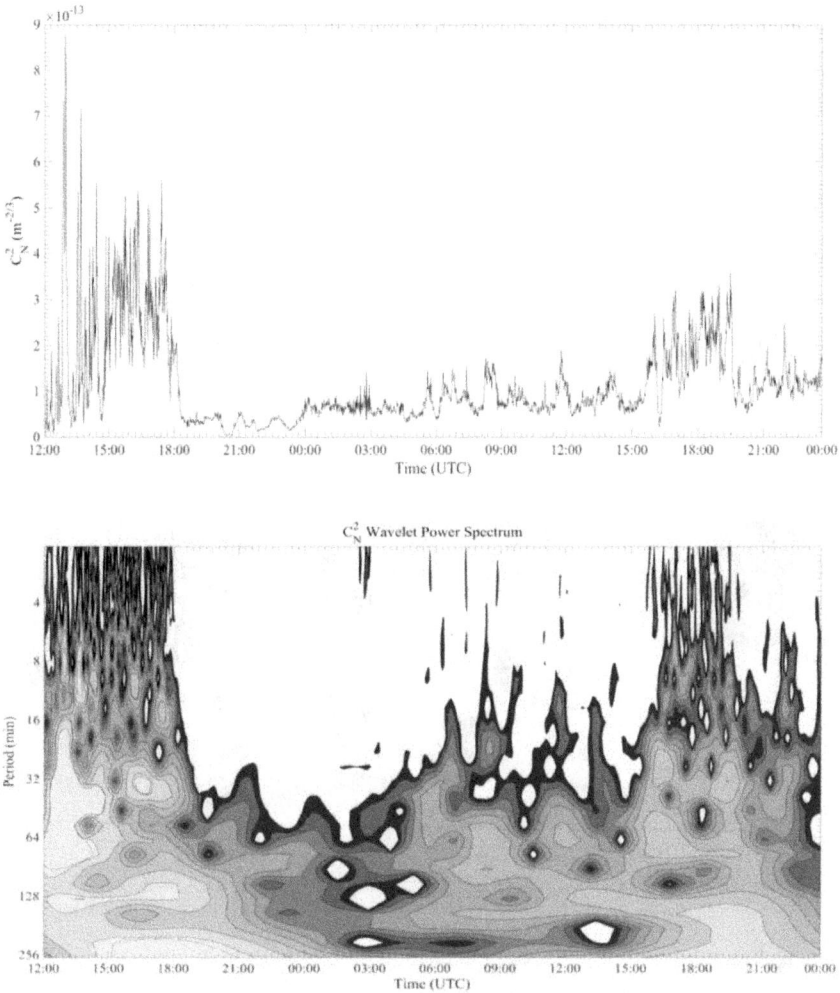

Figure 16. Case I. Turbulence of refractive index $C_N 2$ (m$^{-2/3}$) obtained by LAS. Top panel is the time series of 1 min time integration at 125-Hz laser pulse repetition frequency. Bottom panel is the multiresolution Morlet wavelet spectrogram.

flow is interrupted by introduction of eddies at the period scale of 16–64 min and in response to changes in signal aloft evidenced by C_T^2 in **Figure 11** that shows localized turbulence on shorter time scales than previous Case I. This sequence of flow events resulted in overall negative heat flux excursions ranging from −20 to −40 Wm^{-2} during the first pulse and from −10 to −20 Wm^{-2} in the second event. The spectral feature on the scale of 16–64 min shows turbulence in the flow that could be originated by a combination of variable longitudinal scales aiming the drainage flow and vertical coupling to the ABL where localized shear turbulence of C_T^2 breaks down propagating to the surface. As indicated earlier, each time the drainage flows disrupted the basin, large eddies aiming the flow momentum break through the stratification of the

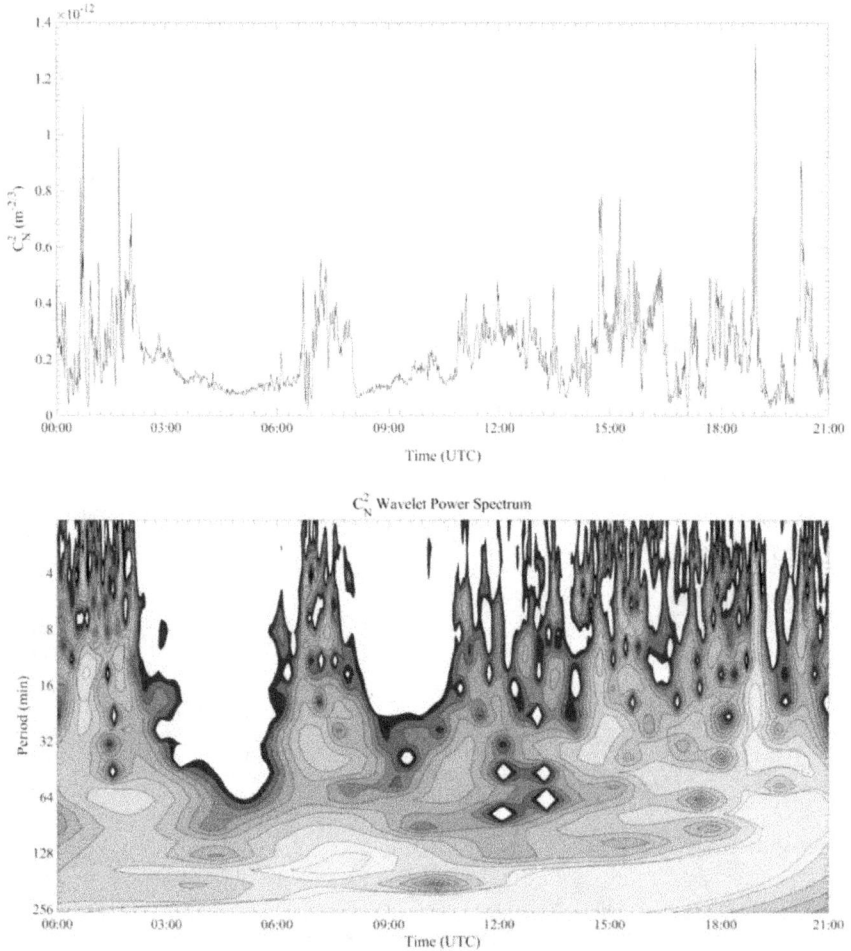

Figure 17. Case II: Turbulence of refractive index C_T^2 (m$^{-2/3}$) obtained by LAS. Top panel is the time series of 1-min time integration at 125-Hz laser pulse repetition frequency. Bottom panel is the multiresolution Morlet wavelet spectrogram.

basin, enabling vertical mixing and therefore allowing cascading turbulence from deep shear layers ~120 m (see **Figure 11**) propagating down to the surface (**Figure 15**).

Finally, for Case III, computed fluxes based on H_{LAS} and H_{EC} diverge in the presence of the drainage flow. For this period, the average values of H_{EC} and H_{LAS} are −20 and −40 Wm^{-2}, respectively. H_{LAS} is systematically higher than H_{EC} throughout the period (03:00 to 18:00 UTC), and this difference is larger than the statistical fluctuation and random errors in EC methodology. The drainage flow enters the basin in the period of time from 9:00 to 15:00, and both H_{LAS} and H_{EC} measurements peak to similar values. The time series of C_N^2 in **Figure 18** top panel displays a large variation in the turbulence of refractive index at the beginning of the drainage flow (00:00 to 4:00 UTC) and at the end when the flow ceased in the period (18:00 to 21:00 UTC).

While the drainage flow is present in the basin, the general trend in C_N^2 is to reduce intensity of turbulence as high-frequency eddies aim the ASL turbulent regime with some fluctuation in the order of 20 min, which explain the difference observed in the heat flux comparison. However, analyzing the vertical structure of the flow, it can be verified that the wind profile experiences shear in layers from 20 to 40 m and at 80 to 160 m. In this case, the periodogram in **Figure 18** bottom panel shows little activity for long periods as the drainage flow evolves in the basin with the exception of two instances when the drainage flow develops localized turbulence in upper levels before 12:00 and between 12:00 and 15:00 UTC, approximately. In these two occasions, it is seen a deposition of turbulent energy in the spectrum on the ~32 min period coincidental with the occurrence of localized upper level ABL C_T^2 turbulence (see **Figure 14**).

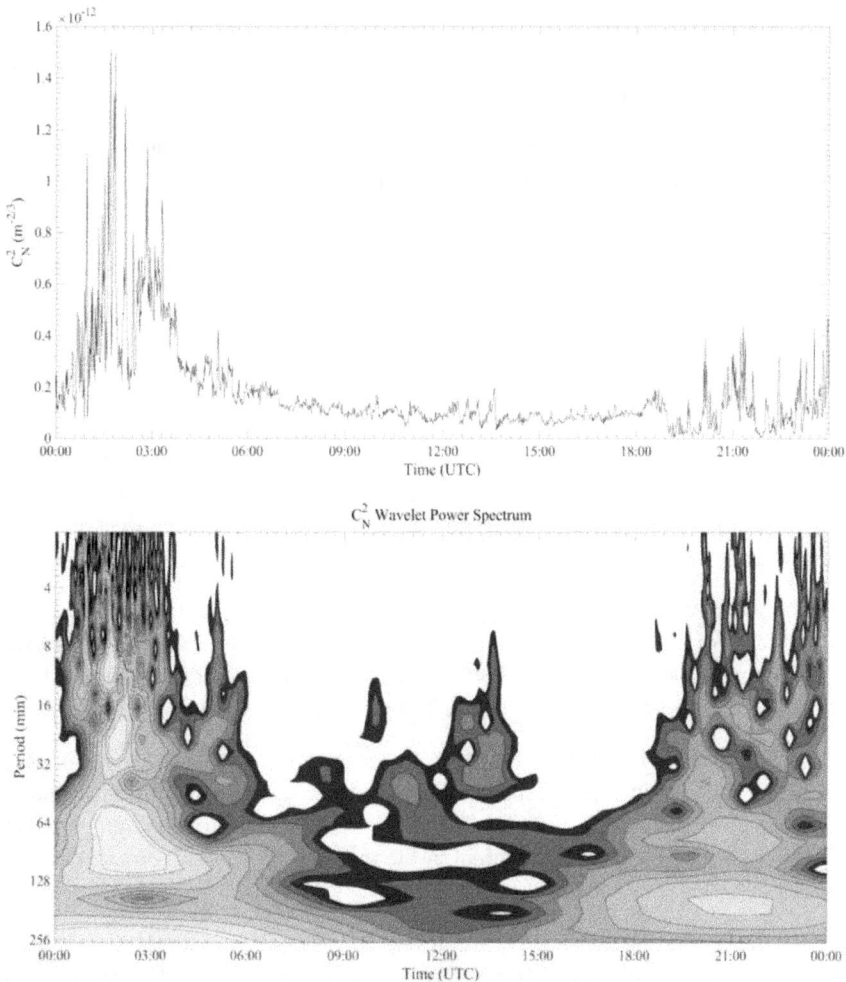

Figure 18. Case III: Turbulence of refractive index obtained C_N^2 by LAS. Top panel is the time series of 1 min time integration at 125 Hz laser pulse repetition frequency. Bottom panel is the multiresolution Morlet wavelet spectrogram.

6. Conclusions

As a summary, this chapter presents an analysis of atmospheric turbulence based on field experiments including instruments for the ASL and the ABL. The objective of Wi-BLEx was to provide a new data set illustrating the dynamic and turbulence regime of a small scale flow penetrating a high latitude basin during extreme winter conditions. The study focuses on quantifying the temporal and spatial aspects of the developing turbulent structures in the ABL and in the ASL and their possible interaction in the conditions high-latitude polar atmospheres. Data sets from Wi-BLEx observations clearly demonstrate the influence of shallow cold drainage flows in the surface turbulent fluxes and the occurrence of large-eddy spectral structures of turbulence during winters in polar regions. Of particular importance in this experiment is the use of high-frequency optical scintillometer to determine the presence of large-eddy turbulence in the ASL that would have been difficult to resolve by sonic anemometers alone in particular when nonstationary flows are under analysis [32].

In this work, three cases were selected for analysis and discussion. The selected cases develop different temporal behavior, surface signatures, and turbulence patterns throughout the vertical structure of the ABL. Of special interest here is the notion that a small scale shallow cold flow entering the basin introduces surface mixing and localized areas of turbulence within the vertical structure of the ABL. Moreover, based on tower observations, it was verified that after drainage flow ceased, the surface layer rapidly re-stratifies in response to the outstanding radiation cooling rate of winter polar atmospheres. In all cases, the drainage flow observed in the vertical by Doppler sodar is represented by wind speed profile that has been observed previously under similar flow dimensions [9, 30, 33].

It is important to note that during winter, the absence of shortwave incoming radiation sets the surface radiation budget to low levels in absolute terms and mostly depending on long-wave net radiation. Therefore, heat fluxes resulting from small scale dynamic processes are limited in absolute values up to 20–30 Wm^{-2} with episodic events of 40 Wm^{-2}.

The cooling effectiveness of the drainage flow was the stronger in the intermittent Case II from all cases but only for short periods of time. A temperature drop of 11° C was verified with an average heat flux of −40 Wm^{-2} during the first intermittent pulse, while the rest of the intermittent events exhibited a lower negative average heat flux in the range of −10 and −5 Wm^{-2}.

In conclusion, during winter, the interaction between polar atmospheres and landscapes combined with the presence of specific synoptic meteorological configurations clearly evidence the possibility of an inter-valley density flow impacting the surface energy balance at a regional scale. This mechanism was evident in the three analyzed cases but also throughout the Wi-BLEx data sets.

Large scale synoptic flows play an important role at regional level to onset the occurrence of drainage flow. Based on the analyzed information, the mode in which the drainage flows develop sustained or intermittent has a combined synoptic and topographic dependence. However, the spatial and temporal scales of the resulting drainage flow depend upon the actual flow-basin dynamic, turbulent, and radiative conditions.

All cases occurred under a surface high pressure forcing in the region. This synoptic meteorological feature is important because it normally comes with a weak pressure gradient force characterized therefore by weak horizontal winds and clear skies, strongly driving the radiative cooling in the basin. During the late winter, Case III, because of the diurnal effect of solar radiation, the study fundamentally differs from the central-winter cases I and II. The drainage event under analysis developed during the night of March 5, 2011 and the day of March 6, 2011. The flow dynamic setting for this case resembles the cases collected during Wi-BLEx during the late winter, in particular for the period March 1 to 10, 2011.

Finally, the three cases described in this study can be summarized as: Case I characterized by a sustained flow lasting ~18 h in the central part of the winter; Case II is characterized by intermittent flow pulses occurred lasting less than ~3 h in the central part of the winter; and Case III resulted a sustained flow lasting for about 9.5 h in late winter. Altogether surface and vertical observations demonstrated that localized turbulence in the ABL depth, as measured in the terms of C_T^2, resulted from an increased flow speed in the ABL that impacted the ASL turbulent regime. In fact, large-eddy dynamic during the flow irruption introduced mixing in the ASL erasing stratification and enabling downward propagation of shear induced thermal turbulence in the ABL to down to the ASL. This is demonstrated by an enhancement of turbulence energy deposition in the low frequency range measured by optical scintillometry.

Acknowledgements

Wi-BLEx was supported by the Air Quality Office of the Fairbanks North Star Borough and by funding from the Department of Environmental Conservation of Alaska. Instrumental support is also recognized from the US Eielson Air Force Base in Alaska. During the Wi-BLEx observational field campaign, both authors were supported by the Geophysical Institute and the College of Natural Science and Mathematics of the University of Alaska Fairbanks. Support for the Chapter Book publication was granted form the Office of the Science of the Vice-Chancellor for Research Prof. Larry Hinzman, University of Alaska Fairbanks.

Conflict of Interest

The authors declare no conflict of interest.

Author details

John A. Mayfield and Gilberto J. Fochesatto*

*Address all correspondence to: gjfochesatto@alaska.edu

Department of Atmospheric Sciences, Geophysical Institute and College of Natural Science and Mathematics, University of Alaska Fairbanks, Fairbanks, Alaska, USA

References

[1] Mayfield JA, Fochesatto GJ. The layered structure of the winter atmospheric boundary layer in the interior of Alaska. Journal of Applied Meteorology and Climatology. 2013;**52**:953-973

[2] Malingowski J, Atkinson D, Fochesatto GJ, Cherry J, Stevens E. An observational study of radiation temperature inversions in Fairbanks, Alaska. Polar Science. 2014;**8**(1):24-39

[3] Bowling SA, Ohtake T, Benson CS. Winter pressure systems and ice fog in Fairbanks, Alaska. Journal of Applied Meteorology. 1968;**7**:961-968

[4] Beran D, Hooke W, Clifford S. Acoustic echo-sounding techniques and their application to gravity-wave, turbulence, and stability studies. Boundary-Layer Meteorology. 1973;**4**:133-153

[5] Holmgren B, Spears L, Wilson C, Benson CS. Acoustic soundings of the Fairbanks temperature inversions. In: Weller G, Bowling SA, editors. Climate of the Arctic: Proceedings of the AAAS-AMS Conference, Fairbanks, Alaska. Geophysical Institute, University of Alaska; 1975, 1973. pp. 293-306

[6] Brown EH, Hall FF Jr. Advances in atmospheric acoustics. Reviews of Geophysics and Space Physics. 1978;**16**:47-110

[7] Fochesatto GJ, Mayfield JA, Gruber MA, Starkenburg D, Conner J. Occurrence of shallow cold flows in the winter atmospheric boundary layer of interior of Alaska. Meteorology and Atmospheric Physics. 2013. DOI: 10.1007/s00703-013-0274-4

[8] Clements WE, Archuleta JA, Hoard D. Mean structure of nocturnal drainage flow in a deep valley. Journal of Applied Meteorology. 1989;**28**:457-462

[9] Doran JC, Horst TW. Observations and models of simple nocturnal slope flows. Journal of the Atmospheric Sciences. 1983;**40**:708-717

[10] Benson CS. Ice Fog: Low Temperature Air Pollution, Defined with Fairbanks, Alaska as Type Locality. College, Alaska: University of Alaska Fairbanks, Geophysical Institute; 1965. p. 134

[11] Wendler G, Jayaweera K. Some measurements of the development of the surface inversion in Central Alaska during winter. Pure and Applied Geophysics. 1972;**99**:209-221

[12] Mölders N, Kramm G. A case study on wintertime inversions in interior Alaska with WRF. Atmospheric Research. 2010;**95**:314-332

[13] Benson CS. Ice fog. Weather. 1970;**25**:11-18

[14] Benson CS. Ice Fog: Low Temperature Air Pollution. CRREL Research Report 121; 1970

[15] Benson CS, Weller G. A Study of Low-Level Winds in the Vicinity of Fairbanks, Alaska. Report to Earth Resources Co. Geophysical Institute, University of Alaska; 1970

[16] Starkenburg D, Metzger S, Fochesatto GJ, Alfieri J, Gens R, Prakash A, et al. Assessment of de-spiking methods for turbulent flux computations in high latitude forest

canopies using sonic anemometers. Journal of Atmospheric and Oceanic Technology. 2016;**33**:2001-2013. DOI: 10.1175/JTECH-D-15-0154.1

[17] Kaimal JC, Finnigan JJ. Atmospheric Boundary Layer Flows: Their Structure and Measurement. Oxford University Press; 1994

[18] Wilczak J, Oncley S, Stage S. Sonic anemometer tilt correction algorithms. Boundary-Layer Meteorology. 2001;**99**:127-150

[19] Vickers D, Mahrt L. Quality control and flux sampling problems for tower and aircraft data. Journal of Atmospheric and Oceanic Technology. 1997;**14**:512-526

[20] Lee X, Massman W, Law B. Editors Handbook of Micrometeorology. A Guide for Surface Flux Measurements and Analysis. Dordrecht, Netherlands: Kluwer Academic Publishers; 2004

[21] Kleissl J, Gomez JSH, Hong SH, Hendrickx JMH, Rahn T, Defoor WL. Large aperture scintillometer intercomparison study. Boundary-Layer Meteorology. 2008;**128**:133-150

[22] De Bruin HAR, Meijninger WML, Smedman AS, Magnusson M. Displaced-beam small aperture scintillometer test. Part I: The Wintex data-set. Boundary-Layer Meteorology. 2002;**105**:129-148

[23] Gruber MA, Fochesatto GJ. A new sensitivity analysis and solution method for scintillometer measurements of area-average turbulent fluxes. Boundary-Layer Meteorology. 2013;**149**:65-83. DOI: 10.1007/s10546-013-9835-9

[24] Gruber MA, Fochesatto GJ, Hartogensis OK, Lysy M. Functional derivatives applied to error propagation of uncertainties in topography to large-aperture scintillometer-derived heat fluxes. Atmospheric Measurement Techniques. 2014;**7**:2361-2371. DOI: 10.5194/amt-7-2361-2014

[25] Fochesatto GJ. Methodology for determining multilayered temperature inversions. Atmospheric Measurement Techniques. 2015;**8**:2051-2060. DOI: 10.5194/amt-8-2051-2015

[26] Neff WD. An observational and numerical study of the atmospheric boundary layer overlying the east antarctic ice sheet. Ph.D. Thesis. Boulder, Colorado: University of Colorado; 1980

[27] Neff WD. Observations of complex terrain flows using acoustic sounders: Echo interpretation. Boundary-Layer Meteorology. 1988;**1988**(40):363-392

[28] Neff WD, King CW. Observations of complex-terrain flows using acoustic sounders–Experiments, topography, and winds. Boundary-Layer Meteorology. 1987;**40**:363-392

[29] Hootman BW, Blumen W. Analysis of nighttime drainage winds in boulder, Colorado during 1980. Monthly Weather Review. 1983;**111**:1052-1061

[30] Sakiyama SK. Drainage flow characteristics and inversion breakup in two Alberta Mountain valleys. Journal of Applied Meteorology. 1990;**29**:1015-1030

[31] Drobinski P, Brown RA, Flamant PH, Pelon J. Evidence of organized large Eddies by ground-based Doppler Lidar, sonic anemometer and sodar. Boundary-Layer Meteorology. 1998;**88**:343-361

[32] Mahrt L. The near-calm stable boundary layer. Boundary-Layer Meteorology. 2011; **140**:343-360

[33] Horst TW, Doran JC. Nocturnal drainage flow on simple slopes. Boundary-Layer Meteorology. 1986;**34**:263-286